FSA Subject Test

Mathematics Grade 5

Student Practice Workbook

+ Two Full-Length Florida FSA Math Tests

Math Notion

www.MathNotion.com

FSA Subject Test Mathematics Grade 5

FSA Subject Test Mathematics Grade 5

Published in the United State of America By

The Math Notion

Web: WWW.MathNotion.com

Email: info@MathNotion.com

Copyright © 2021 by the Math Notion. All rights reserved. No part of this publication may be reproduced, stored in a retrieval system, or transmitted in any form or by any means, electronic, mechanical, photocopying, recording, scanning, or otherwise, except as permitted under Section 107 or 108 of the 1976 United States Copyright Ac, without permission of the author.

All inquiries should be addressed to the Math Notion.

ISBN: 978-1-63620-081-1

The Math Notion

Michael Smith has been a math instructor for over a decade now. He launched the Math Notion. Since 2006, we have devoted our time to both teaching and developing exceptional math learning materials. As a test prep company, we have worked with thousands of students. We have used the feedback of our students to develop a unique study program that can be used by students to drastically improve their math scores fast and effectively. We have more than a thousand Math learning books including:

- **SAT Math Prep**
- **ACT Math Prep**
- **SSAT/ISEE Math Prep**
- **Mathematics Prep Grade 3 to 8**
- **Common Core Math Prep**
- **many Math Education Workbooks, Study Guides, Practice and Exercise Books**

As an experienced Math test preparation company, we have helped many students raise their standardized test scores—and attend the colleges of their dreams: We tutor online and in person, we teach students in large groups, and we provide training materials and textbooks through our website and through Amazon.

You can contact us via email at:

info@Mathnotion.com

FSA Subject Test Mathematics Grade 5

Get the Targeted Practice You Need to Ace the FSA Math Test!

FSA Subject Test Mathematics Grade 5 includes easy-to-follow instructions, helpful examples, and plenty of math practice problems to assist students to master each concept, brush up their problem-solving skills, and create confidence.

The FSA math practice book provides numerous opportunities to evaluate basic skills along with abundant remediation and intervention activities. It is a skill that permits you to quickly master intricate information and produce better leads in less time.

Students can boost their test-taking skills by taking the book's two practice FSA Math exams. All test questions answered and explained in detail.

Important Features of the 5th grade FSA Math Book:

- A **complete review** of FSA math test topics,
- Over 2,500 practice problems covering all topics tested,
- The most important concepts you need to know,
- Clear and concise, easy-to-follow sections,
- Well designed for enhanced learning and interest,
- Hands-on experience with all question types,
- **2 full-length practice tests** with detailed answer explanations,
- Cost-Effective Pricing,

Powerful math exercises to help you avoid traps and pacing yourself to beat the Florida FSA test. Students will gain valuable experience and raise their confidence by taking 5th grade math practice tests, learning about test structure, and gaining a deeper understanding of what is tested on the FSA math grade 5. If ever there was a book to respond to the pressure to increase students' test scores, this is it.

WWW.MathNotion.COM

… So Much More Online!

- ✓ FREE Math Lessons
- ✓ More Math Learning Books!
- ✓ Mathematics Worksheets
- ✓ Online Math Tutors

For a PDF Version of This Book

Please Visit WWW.MathNotion.com

FSA Subject Test Mathematics Grade 5

Contents

Chapter 1 : Place Values and Number Sense 11
    Place Values 12
    Comparing and Ordering Numbers 13
    Numbers in Word Form 14
    Roman Numerals 15
    Rounding Numbers 16
    Odd or Even 17
    Repeating Patterns 18
    Growing Patterns 19
    Patterns: Numbers 20
    Answers of Worksheets 21

Chapter 2 : Whole Number Operations 23
    Adding Whole Numbers 24
    Subtracting Whole Numbers 25
    Multiplying Whole Numbers 26
    Dividing Hundreds 27
    Long Division by Two Digits 28
    Division with Remainders 28
    Rounding Whole Numbers 29
    Whole Number Estimation 30
    Answers of Worksheets 31

Chapter 3 : Number Theory 33
    Factoring Numbers 34
    Prime Factorization 34
    Divisibility Rules 35
    Greatest Common Factor 36
    Least Common Multiple 36
    Answers of Worksheets 37

Chapter 4 : Fractions and Mixed Numbers 39
    Simplifying Fractions 40
    Like Denominators 41
    Compare Fractions with Like Denominators 43
    More Than Two Fractions with Like Denominators 44
    Unlike Denominators 45
    Ordering Fractions 47

WWW.MathNotion.Com

FSA Subject Test Mathematics Grade 5

    Denominators of 10, 100, and 1000 48
    Fractions to Mixed Numbers 50
    Mixed Numbers to Fractions 51
    Add and Subtract Mixed Numbers 52
    Answers of Worksheets 53

Chapter 5 : Decimals **57**
    Adding and Subtracting Decimals 58
    Multiplying and Dividing Decimals 59
    Rounding Decimals 60
    Comparing Decimals 61
    Answers of Worksheets 62

Chapter 6 : Ratios and Rates **63**
    Simplifying Ratios 64
    Writing Ratios 64
    Create a Proportion 65
    Proportional Ratios 65
    Similar Figures 66
    Word Problems 67
    Answers of Worksheets 69

Chapter 7 : Measurement **71**
    Reference Measurement Units 72
    Metric Length Units 73
    Customary Length Units 73
    Metric Capacity Units 74
    Customary Capacity Units 74
    Metric Weight and Mass Units 75
    Customary Weight and Mass Units 75
    Temperature Units 76
    Time 77
    Money Amounts 78
    Money: Word Problems 79
    Answers of Worksheets 80

Chapter 8 : Algebraic Thinking **82**
    Finding Rules 83
    Algebraic Word Problems 84
    Evaluate Expressions 85
    Variables and Expressions 86
    Answers of Worksheets 87

FSA Subject Test Mathematics Grade 5

Chapter 9 : Geometric **89**
   Identifying Angles 90
   Estimate Angle Measurements 91
   Measure Angles with a Protractor 92
   Polygon Names 93
   Classify Triangles 94
   Parallel Sides in Quadrilaterals 95
   Identify Rectangles 96
   Perimeter: Find the Missing Side Lengths 97
   Perimeter and Area of Squares 98
   Perimeter and Area of rectangles 99
   Find the Area or Missing Side Length of a Rectangle 100
   Area and Perimeter: Word Problems 101
   Circumference, Diameter, and Radius 102
   Volume of Cubes and Rectangle Prisms 103
   Answers of Worksheets 104

Chapter 10 : Three-Dimensional Figures **106**
   Identify Three–Dimensional Figures 107
   Count Vertices, Edges, and Faces 108
   Identify Faces of Three–Dimensional Figures 109
   Answers of Worksheets 110

Chapter 11 : Symmetry and Transformations **111**
   Line Segments 112
   Identify Lines of Symmetry 113
   Count Lines of Symmetry 114
   Parallel, Perpendicular and Intersecting Lines 115
   Answers of Worksheets 116

Chapter 12 : Data Graphs, and Statistics **117**
   Mean and Median 118
   Mode and Range 119
   Graph Points on a Coordinate Plane 120
   Bar Graph 121
   Tally and Pictographs 122
   Dot plots 123
   Line Graphs 124
   Stem–And–Leaf Plot 125
   Scatter Plots 126
   Probability Problems 127

FSA Subject Test Mathematics Grade 5

Answers of Worksheets 128
Chapter 13 : FSA Math Practice Tests **131**
    FSA GRADE 5 MAHEMATICS REFRENCE MATERIALS 133
    FSA Practice Test 1 135
        Session 1 136
        Session 2 142
    FSA Practice Test 2 149
        Session 1 150
        Session 2 156
Chapter 14 : Answers and Explanations **163**
    Answer Key 163
    Practice Test 1 165
    Practice Test 2 169

FSA Subject Test Mathematics Grade 5

Chapter 1 : Place Values and Number Sense

Topics that you'll learn in this chapter:

- ✓ Place Values,
- ✓ Compare and Ordering Numbers,
- ✓ Numbers in Word Form,
- ✓ Roman Numerals,
- ✓ Rounding Numbers,
- ✓ Odd or Even,
- ✓ Repeating Patterns,
- ✓ Growing Patterns,
- ✓ Patterns: Numbers,

FSA Subject Test Mathematics Grade 5

Place Values

✏️ Write numbers in expanded form.

1) Sixty–two   ___ + ___

2) fifty–six   ___ + ___

3) thirty–one   ___ + ___

4) forty–five   ___ + ___

5) twenty–eight   ___ + ___

✏️ Circle the correct choice.

6) The 6 in 56 is in the

        Ones place   tens place   hundreds place

7) The 2 in 25 is in the

        Ones place   tens place   hundreds place

8) The 9 in 918 is in the

        Ones place   tens place   hundreds place

9) The 3 in 537 is in the

        Ones place   tens place   hundreds place

10) The 9 in 289 is in the

        Ones place   tens place   hundreds place

FSA Subject Test Mathematics Grade 5

Comparing and Ordering Numbers

✏ Use less than, equal to or greater than.

1) 31 _____ 33
2) 57 _____ 49
3) 92 _____ 88
4) 76 _____ 67
5) 43 _____ 43
6) 54 _____ 46
7) 97 _____ 88

8) 42 _____ 36
9) 55 _____ 55
10) 57 _____ 75
11) 28 _____ 38
12) 19 _____ 15
13) 82 _____ 90
14) 78 _____ 84

✏ Order each set numbers from least to greatest.

15) – 18, – 22, 28, – 17, 4   ___, ___, ___, ___, ___, ___

16) 19, –36, 11, – 12, 5   ___, ___, ___, ___, ___, ___

17) 27, – 56, 20, 1, – 27   ___, ___, ___, ___, ___, ___

18) 26, – 96, 2, – 26, 87, –75   ___, ___, ___, ___, ___, ___

19) –10, –71, 70, –26, –59, –39   ___, ___, ___, ___, ___, ___

20) 88, 4, 38, 7, 78, 9   ___, ___, ___, ___, ___, ___

21) 84, 14, 24, 0, 35, 22   ___, ___, ___, ___, ___, ___

WWW.MathNotion.Com

Numbers in Word Form

✎ Write each number in words.

1) 372 _____

2) 605 _____

3) 550 _____

4) 351 _____

5) 793 _____

6) 647 _____

7) 3,219 _____

8) 5,326 _____

9) 2,842 _____

10) 4,691 _____

11) 5,531 _____

12) 7,360 _____

13) 2,532 _____

14) 8,014 _____

15) 11,242 _____

Roman Numerals

✍ Write in Romans numerals.

1	I	11	XI	21	XXI
2	II	12	XII	22	XXII
3	III	13	XIII	23	XXIII
4	IV	14	XIV	24	XXIV
5	V	15	XV	25	XXV
6	VI	16	XVI	26	XXVI
7	VII	17	XVII	27	XXVII
8	VIII	18	XVIII	28	XXVIII
9	IX	19	XIX	29	XXIX
10	X	20	XX	30	XXX

1) 11 _____ 2) 21 _____

3) 24 _____ 4) 16 _____

5) 27 _____ 6) 29 _____

7) 12 _____ 8) 28 _____

9) 15 _____ 10) 20 _____

11) Add 16 + 14 and write in Roman numerals. _____

12) Subtract 34 − 5 and write in Roman numerals. _____

FSA Subject Test Mathematics Grade 5

Rounding Numbers

✎ Round each number to the underlined place value.

1) 3,<u>7</u>93

2) 3,<u>8</u>76

3) 3,4<u>5</u>2

4) 7,1<u>9</u>3

5) 5,2<u>7</u>8

6) 1,4<u>7</u>7

7) 8,<u>3</u>13

8) 24.<u>6</u>8

9) 8<u>4</u>.92

10) 71.<u>3</u>4

11) 66<u>4</u>.7

12) <u>9</u>,135

13) 15.3<u>8</u>1

14) 4,<u>5</u>21

15) 3<u>6</u>.50

16) 4,<u>8</u>19

17) 6,6<u>8</u>5

18) 2,5<u>3</u>8

19) 73.<u>6</u>2

20) 16,<u>5</u>27

21) 2<u>9</u>.720

22) 12,3<u>6</u>6

23) 31,<u>7</u>29

24) 7,8<u>3</u>8

FSA Subject Test Mathematics Grade 5

Odd or Even

🖎 Identify whether each number is even or odd.

1) 18 _____   7) 80 _____

2) 27 _____   8) 53 _____

3) 21 _____   9) 58 _____

4) 17 _____   10) 98 _____

5) 67 _____   11) 49 _____

6) 76 _____   12) 113 _____

🖎 Circle the even number in each group.

13) 52, 11, 35, 73, 5, 29   15) 33, 45, 86, 59, 63, 87

14) 13, 15, 113, 87, 71, 18   16) 55, 32, 79, 51, 21, 83

🖎 Circle the odd number in each group.

17) 54, 36, 48, 76, 71, 100   19) 58, 92, 25, 78, 76, 50

18) 32, 56, 40, 74, 98, 67   20) 89, 12, 88, 42, 48, 120

FSA Subject Test Mathematics Grade 5

Repeating Patterns

✎ Circle the picture that comes next in each picture pattern.

1)

2)

3)

4)

5)

Growing Patterns

✎ Draw the picture that comes next in each growing pattern.

1)

2)

3)

4)

5)

FSA Subject Test Mathematics Grade 5

Patterns: Numbers

✏ Write the numbers that come next.

1) 2, 5, 8, 11,   ____, ____, ____, ____

2) 10, 15, 20, 25,   ____, ____, ____, ____

3) 4, 8, 12, 16,   ____, ____, ____, ____

4) 7, 17, 27, 37,   ____, ____, ____, ____

5) 5, 12, 19, 26,   ____, ____, ____, ____

6) 8, 16, 24, 32, 40,   ____, ____, ____, ____

✏ Write the next three numbers in each counting sequence.

1) −31, −19, −7, _____, _____, _____, _____

2) 541, 526, 511, _____, _____, _____, _____

3) 14, 34, _____, _____, 94, _____

4) 21, 29, _____, _____, _____

5) 89, 78, _____, _____, _____

6) 95, 82, 69, _____, _____, _____

7) 198, 166, 134, _____, _____, _____

8) What are the next three numbers in this counting sequence?

   1870, 1970, 2070, _____, _____, _____

9) What is the fourth number in this counting sequence?

   8, 14, 20, _____

FSA Subject Test Mathematics Grade 5

Answers of Worksheets

Place Values

1) 60 + 2
2) 50 + 6
3) 30 + 1
4) 40 + 5
5) 20 + 8
6) ones place
7) tens place
8) hundreds place
9) tens place
10) ones place

Comparing and Ordering Numbers

1) 31 less than 33
2) 57 greater than 49
3) 92 greater than 88
4) 76 greater than 67
5) 43 equals to 43
6) 54 greater than 46
7) 97 greater than 88
8) 42 greater than 36
9) 55 equals to 55
10) 57 less than 75
11) 28 less than 38
12) 19 greater than 15
13) 82 less than 90
14) 78 less than 84
15) –22, –18, –17, 4, 28
16) –36, –12, 5, 11, 19
17) –56, –27, 1, 20, 27
18) –96, –75, –26, 2, 26, 87
19) –71, –59, –39, –26, –10, 70
20) 4, 7, 9, 38, 78, 88
21) 0, 14, 22, 24, 35, 84

Numbers in Word Form

1) three hundred seventy-two
2) six hundred five
3) five hundred fifty
4) three hundred fifty-one
5) seven hundred ninety-three
6) six hundred forty-seven
7) three thousand, two hundred nineteen
8) five thousand, three hundred twenty-six
9) two thousand, eight hundred forty-two
10) four thousand, six hundred ninety-one
11) five thousand, five hundred thirty-one
12) seven thousand, three hundred sixty
13) two thousand, five hundred thirty-two
14) eight thousand, fourteen
15) eleven thousand, two hundred forty-two

Roman Numerals

1) XI
2) XXI
3) XXIV
4) XVI
5) XXVII
6) XXIX
7) XII
8) XXVIII
9) XV
10) XX
11) XXX
12) XXIX

Rounding Numbers

1) 4,000
2) 4,000
3) 3,450
4) 7,190
5) 5,280
6) 1,480
7) 8,300
8) 24.70
9) 85.00
10) 71.30
11) 665.00
12) 9,000

FSA Subject Test Mathematics Grade 5

13) 15.380	16) 4,800	19) 73.60	22) 12,370
14) 4,500	17) 6,700	20) 16,500	23) 31,700
15) 37.00	18) 2,540	21) 30.00	24) 7,840

Odd or Even

1) even	6) even	11) odd	16) 32
2) odd	7) even	12) odd	17) 71
3) odd	8) odd	13) 52	18) 67
4) odd	9) even	14) 18	19) 25
5) odd	10) even	15) 86	20) 89

Repeating pattern

1)
2)
3)
4) (heart)
5) (heart)

Growing patterns

1)
2)
3)
4)
5)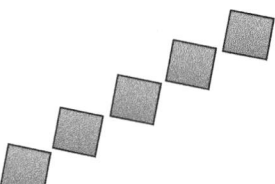

Patterns: Numbers

1) 2, 5, 8, 11, 14, 17, 20, 23
2) 10, 15, 20, 25, 30, 35, 40, 45
3) 4, 8, 12, 16, 20, 24, 28, 32
4) 7, 17, 27, 37, 47, 57, 67, 77
5) 5, 12, 19, 26, 33, 40, 47, 54
6) 8, 16, 24, 32, 40, 48, 56, 64

Patterns

1) 5, 17, 29, 41
2) 496, 481, 466, 451
3) 14, 34, 54, 74, 94, 114
4) 37, 45, 53
5) 67, 56, 45
6) 56, 43, 30
7) 102, 70, 38
8) 2170, 2270, 2370
9) 26

WWW.MathNotion.Com

Chapter 2 : Whole Number Operations

Topics that you'll learn in this chapter:

- ✓ Adding Whole Numbers,
- ✓ Subtracting Whole Numbers,
- ✓ Multiplying Whole Numbers,
- ✓ Dividing Hundreds,
- ✓ Long Division by One Digit,
- ✓ Division with Remainders,
- ✓ Rounding Whole Numbers,
- ✓ Whole Number Estimation,

FSA Subject Test Mathematics Grade 5

Adding Whole Numbers

✎ Add.

1) 5,763
   + 8,238
   _____

2) 6,834
   + 4,998
   _____

3) 3,548
   + 5,693
   _____

4) 2,769
   + 8,872
   _____

5) 3,196
   + 2,936
   _____

6) 7,009
   + 4,992
   _____

✎ Find the missing numbers.

7) 3,468 + ___ = 4,102

8) 840 + 2,360 = ___

9) 5,200 + ___ = 7,980

10) 631 + ___ = 2,007

11) ___ + 803 = 3,945

12) ___ + 2,156 = 5,922

13) David sells gems. He finds a diamond in Istanbul and buys it for $4,795. Then, he flies to Cairo and purchases a bigger diamond for the bargain price of $9,633. How much does David spend on the two diamonds? _____

WWW.MathNotion.Com

FSA Subject Test Mathematics Grade 5

Subtracting Whole Numbers

✎ Subtract.

1) 10,512 − 4,411

2) 5,204 − 3,679

3) 8,520 − 6,483

4) 8,001 − 5,224

5) 11,916 − 8,711

6) 5,005 − 2,008

✎ Find the missing number.

7) 5,263 − ___ = 2,367

8) 7,198 − ___ = 4,742

9) 8,928 − 3,764 = ___

10) 6,511 − ___ = 3,759

11) 7,003 − 5,489 = ___

12) 8,800 − 5,995 = ___

13) Jackson had $7,189 invested in the stock market until he lost $3,793 on those investments. How much money does he have in the stock market now?
_____

WWW.MathNotion.Com

Multiplying Whole Numbers

✎ Find the answers.

1) 2,200 × 31

2) 3,200 × 22

3) 5,790 × 5

4) 5,220 × 3

5) 6,911 × 3

6) 1,998 × 40

7) 2,893 × 5.5

8) 2,254 × 3.5

9) 4,372 × 4.8

10) 3,984 × 2.75

11) 4,900 × 2.5

12) 8,200 × 4.5

Dividing Hundreds

✎ Find answers.

1) 4,440 ÷ 400

2) 1,600 ÷ 40

3) 9,990 ÷ 90

4) 4,200 ÷ 60

5) 6,400 ÷ 8,000

6) 2,700 ÷ 30

7) 3,333 ÷ 30

8) 558 ÷ 45

9) 2,278 ÷ 85

10) 1,683 ÷ 55

11) 1,582 ÷ 35

12) 9,000 ÷ 600

13) 1,000 ÷ 2,500

14) 44.8 ÷ 20

15) 6,800 ÷ 400

16) 1,500 ÷ 5,000

17) 36.60 ÷ 120

18) 7,700 ÷ 700

19) 5,400 ÷ 600

20) 8,000 ÷ 160

21) 18,000 ÷ 9,000

22) 42,000 ÷ 30

23) 480 ÷ 40

24) 63,000 ÷ 900

FSA Subject Test Mathematics Grade 5

Long Division by Two Digits

✎ Find the quotient.

1) 18)576

2) 14)952

3) 21)588

4) 23)299

5) 44)748

6) 26)234

7) 16)496

8) 29)1,479

9) 54)1,080

10) 41)1,476

11) 53)2,491

12) 60)2,880

13) 32)2,912

14) 77)8,393

15) 85)3,740

16) 57)4,617

17) 50)9,200

18) 25)15,400

Division with Remainders

✎ Find the quotient with remainder.

1) 14)715

2) 16)2,750

3) 27)4,603

4) 58)2,554

5) 42)7,732

6) 63)6,737

7) 71)9,036

8) 65)8,624

9) 35)5,705

10) 92)13,161

11) 46)12,214

12) 69)42,482

13) 85)6,858

14) 87)34,304

FSA Subject Test Mathematics Grade 5

Rounding Whole Numbers

🖎 Round each number to the underlined place value.

1) 7,533

2) 9,374

3) 8,883

4) 2,368

5) 5,577

6) 3,381

7) 3,520

8) 9,338

9) 8.581

10) 33.57

11) 51.69

12) 22.138

13) 6,758

14) 11,557

15) 8,838

16) 5.889

17) 1.860

18) 25.070

19) 9.332

20) 49.48

21) 28.89

22) 24,377

23) 52,158

24) 13,883

25) 9,609

26) 17,451

27) 18,768

FSA Subject Test Mathematics Grade 5

Whole Number Estimation

✎ Estimate the sum by rounding each added to the nearest ten.

1) 875 + 325

2) 985 + 1,452

3) 2,424 + 4,128

4) 1,576 + 6,279

5) 1,247 + 3,863

6) 6,746 + 5,121

7) 3,924 + 6,456

8) 1,785 + 7,164

9) 1,458
   + 2,442
   ―――――

10) 5,689
    + 4,151
    ―――――

11) 8,259
    + 4,754
    ―――――

12) 6,788
    + 3,954
    ―――――

13) 9,123
    + 4,455
    ―――――

14) 6,680
    + 5,358
    ―――――

15) 3,165
    + 7,124
    ―――――

16) 8,859
    + 6,452
    ―――――

FSA Subject Test Mathematics Grade 5

Answers of Worksheets

Adding Whole Numbers

1) 14,001
2) 11,832
3) 9,241
4) 11,641
5) 6,132
6) 12,001
7) 634
8) 3,200
9) 2,780
10) 1,376
11) 3,142
12) 3,766
13) $14,428

Subtracting Whole Numbers

1) 6,101
2) 1,525
3) 2,037
4) 2,777
5) 3,205
6) 2,997
7) 2,896
8) 2,456
9) 5,164
10) 2,752
11) 1,514
12) 2,805
13) 3,396

Multiplying Whole Numbers

1) 68,200
2) 70,400
3) 28,950
4) 15,660
5) 20,733
6) 79,920
7) 15,911.5
8) 7,889
9) 20,985.6
10) 10,956
11) 12,250
12) 36,900

Dividing Hundreds

1) 11.1
2) 40
3) 111
4) 70
5) 0.8
6) 90
7) 111.1
8) 12.4
9) 26.8
10) 30.6
11) 45.2
12) 15
13) 0.4
14) 2.24
15) 17
16) 0.3
17) 0.305
18) 11
19) 9
20) 50
21) 2
22) 1,400
23) 12
24) 70

Long Division by Two Digits

1) 32
2) 68
3) 28
4) 13
5) 17
6) 9
7) 31
8) 51
9) 20
10) 36
11) 47
12) 48
13) 91
14) 109
15) 44
16) 81
17) 184
18) 616

WWW.MathNotion.Com

FSA Subject Test Mathematics Grade 5

Division with Remainders

1) 51 R1
2) 171 R14
3) 170 R13
4) 44 R2
5) 184 R4
6) 106 R59
7) 127 R19
8) 132 R44
9) 163 R0
10) 143 R5
11) 265 R24
12) 615 R47
13) 80 R58
14) 394 R26

Rounding Whole Numbers

1) 7,500
2) 9,400
3) 8,880
4) 2,370
5) 5,580
6) 3,380
7) 3,500
8) 9,340
9) 8.60
10) 33.60
11) 51.70
12) 22.100
13) 7,000
14) 11,560
15) 8,840
16) 5.900
17) 1.900
18) 25.100
19) 9.000
20) 49.50
21) 28.90
22) 24,380
23) 52,160
24) 13,880
25) 9,600
26) 17,450
27) 18,800

Whole Number Estimation

1) 1,200
2) 2,440
3) 6,550
4) 7,860
5) 5,110
6) 11,870
7) 10,380
8) 8,950
9) 3,900
10) 9,840
11) 13,010
12) 10,740
13) 13,580
14) 12,040
15) 10,290
16) 15,310

WWW.MathNotion.Com

Chapter 3 : Number Theory

Topics that you'll learn in this chapter:

- ✓ Factoring Numbers,
- ✓ Prime Factorization,
- ✓ Divisibility Rules,
- ✓ Greatest Common Factor,
- ✓ Least Common Multiple,

FSA Subject Test Mathematics Grade 5

Factoring Numbers

✎ List all positive factors of each number.

1) 12	6) 56	11) 27
2) 16	7) 65	12) 63
3) 28	8) 70	13) 72
4) 34	9) 25	14) 15
5) 95	10) 48	15) 80

✎ List the prime factorization for each number.

16) 10	19) 30	22) 55
17) 26	20) 40	23) 78
18) 20	21) 44	24) 96

Prime Factorization

✎ Factor the following numbers to their prime factors.

1) 6	9) 58	17) 69
2) 49	10) 62	18) 76
3) 60	11) 75	19) 86
4) 4	12) 88	20) 92
5) 46	13) 93	21) 99
6) 57	14) 100	22) 77
7) 54	15) 68	23) 90
8) 38	16) 90	24) 74

WWW.MathNotion.Com

Divisibility Rules

✎ Use the divisibility rules to underline the factors of the number.

1) 8   2 3 4 5 6 7 8 9 10

2) 18   2 3 4 5 6 7 8 9 10

3) 55   2 3 4 5 6 7 8 9 10

4) 45   2 3 4 5 6 7 8 9 10

5) 20   2 3 4 5 6 7 8 9 10

6) 9   2 3 4 5 6 7 8 9 10

7) 21   2 3 4 5 6 7 8 9 10

8) 28   2 3 4 5 6 7 8 9 10

9) 36   2 3 4 5 6 7 8 9 10

10) 40 2 3 4 5 6 7 8 9 10

11) 39 2 3 4 5 6 7 8 9 10

12) 51 2 3 4 5 6 7 8 9 10

FSA Subject Test Mathematics Grade 5

Greatest Common Factor

🖎 Find the GCF for each number pair.

1) 25, 15	9) 52, 3	17) 66, 18
2) 8, 18	10) 12, 54	18) 70, 15
3) 14, 28	11) 11, 13	19) 38, 14
4) 18, 32	12) 56, 48	20) 36, 28
5) 15, 45	13) 75, 25	21) 100, 60
6) 22, 33	14) 40, 60	22) 85, 35
7) 19, 21	15) 52, 32	23) 16, 48
8) 27, 72	16) 30, 55	24) 13, 39

Least Common Multiple

🖎 Find the LCM for each number pair.

1) 3, 15	9) 13, 26	17) 13, 2, 26
2) 5, 35	10) 15, 65	18) 18, 6, 24
3) 24, 16	11) 12, 8	19) 9, 12, 15
4) 28, 40	12) 6, 44	20) 7, 12, 4
5) 9, 27	13) 10, 16	21) 5, 15, 16
6) 46, 23	14) 7, 6	22) 13, 4, 26
7) 22, 66	15) 12, 36, 24	23) 3, 14, 5
8) 4, 9	16) 5, 11, 2	24) 32, 8, 3

FSA Subject Test Mathematics Grade 5

Answers of Worksheets

Factoring Numbers

1) 1, 2, 3, 4, 6, 12
2) 1, 2, 4, 8, 16
3) 1, 2, 4, 7, 14, 28
4) 1, 2, 17, 34
5) 1, 5, 19, 95
6) 1, 2, 4, 7, 8, 14, 28, 56
7) 1, 5, 13, 65
8) 1, 2, 5, 7, 10, 14, 35, 70
9) 1, 5, 25
10) 1, 2, 3, 4, 6, 8, 12, 16, 24, 48
11) 1, 3, 9, 27
12) 1, 3, 7, 9, 21, 63
13) 1, 2, 3, 4, 6, 8, 9, 12, 18, 24, 36, 72
14) 1, 3, 5, 15
15) 1, 2, 4, 5, 8, 10, 16, 20, 40, 80
16) 2 × 5
17) 2 × 13
18) 2 × 2 × 5
19) 2 × 3 × 5
20) 2 × 2 × 2 × 5
21) 2 × 2 × 11
22) 5 × 11
23) 2 × 3 × 13
24) 2 × 2 × 2 × 2 × 2 × 3

Prime Factorization

1) 2. 3
2) 7. 7
3) 2. 2. 3. 5
4) 2. 2
5) 2. 23
6) 3. 19
7) 2. 3. 3. 3
8) 2. 19
9) 2. 29
10) 2. 31
11) 3. 5. 5
12) 2. 2. 2. 11
13) 3. 31
14) 2. 2. 5. 5
15) 2. 2. 17
16) 2. 3. 3. 5
17) 3. 23
18) 2. 2. 19
19) 2. 43
20) 2. 2. 23
21) 3. 3. 11
22) 7. 11
23) 2. 3. 3. 5
24) 2. 37

Divisibility Rules

1) 8   <u>2</u> 3 <u>4</u> 5 6 7 <u>8</u> 9 10
2) 18   <u>2</u> <u>3</u> 4 5 <u>6</u> 7 8 <u>9</u> 10
3) 55   2 3 4 <u>5</u> 6 7 8 9 10
4) 45   2 <u>3</u> 4 <u>5</u> 6 7 8 <u>9</u> 10
5) 20   <u>2</u> 3 <u>4</u> <u>5</u> 6 7 8 9 <u>10</u>
6) 9   2 <u>3</u> 4 5 6 7 8 <u>9</u> 10

FSA Subject Test Mathematics Grade 5

7) 21   2 <u>3</u> 4 5 6 <u>7</u> 8 9 10

8) 28   <u>2</u> 3 <u>4</u> 5 6 <u>7</u> 8 9 10

9) 36   <u>2</u> <u>3</u> <u>4</u> 5 <u>6</u> 7 8 <u>9</u> 10

10) 40   <u>2</u> 3 <u>4</u> <u>5</u> 6 7 <u>8</u> 9 <u>10</u>

11) 39   2 <u>3</u> 4 5 6 7 8 9 10

12) 51   2 <u>3</u> 4 5 6 7 8 9 10

Greatest Common Factor

1) 5   7) 1   13) 25   19) 2

2) 2   8) 9   14) 20   20) 4

3) 14   9) 1   15) 4   21) 20

4) 2   10) 6   16) 5   22) 5

5) 15   11) 1   17) 6   23) 16

6) 11   12) 8   18) 5   24) 13

Least Common Multiple

1) 15   7) 66   13) 80   19) 180

2) 35   8) 36   14) 42   20) 84

3) 48   9) 26   15) 72   21) 240

4) 280   10) 195   16) 110   22) 52

5) 27   11) 24   17) 26   23) 210

6) 46   12) 132   18) 72   24) 96

WWW.MathNotion.Com

FSA Subject Test Mathematics Grade 5

Chapter 4 : Fractions and Mixed Numbers

Topics that you'll learn in this chapter:

- ✓ Simplifying Fractions,
- ✓ Like Denominators,
- ✓ Compare Fractions with Like Denominators,
- ✓ More than two Fractions with Like Denominators,
- ✓ Unlike Denominators,
- ✓ Ordering Fractions,
- ✓ Denominators of 10, 100, and 1000,
- ✓ Fractions to Mixed Numbers,
- ✓ Mixed Numbers to Fractions,
- ✓ Add and Subtract Mixed Numbers,

Simplifying Fractions

✎ Simplify the fractions.

1) $\dfrac{44}{84}$

2) $\dfrac{8}{20}$

3) $\dfrac{12}{16}$

4) $\dfrac{4}{24}$

5) $\dfrac{15}{30}$

6) $\dfrac{9}{63}$

7) $\dfrac{4}{14}$

8) $\dfrac{17}{51}$

9) $\dfrac{24}{30}$

10) $\dfrac{5}{35}$

11) $\dfrac{16}{48}$

12) $\dfrac{33}{22}$

13) $\dfrac{45}{63}$

14) $\dfrac{2.4}{3.2}$

15) $\dfrac{12}{60}$

16) $\dfrac{70}{112}$

17) $\dfrac{2.7}{7.2}$

18) $\dfrac{33}{88}$

19) $\dfrac{1.5}{13.5}$

20) $\dfrac{39}{52}$

21) $\dfrac{5}{45}$

22) $\dfrac{2.1}{4.2}$

FSA Subject Test Mathematics Grade 5

Like Denominators

✎ Add fractions.

1) $\dfrac{3}{4}+\dfrac{1}{4}$

2) $\dfrac{1}{5}+\dfrac{4}{5}$

3) $\dfrac{4}{9}+\dfrac{7}{9}$

4) $\dfrac{2}{7}+\dfrac{2}{7}$

5) $\dfrac{5}{13}+\dfrac{2}{13}$

6) $\dfrac{1}{14}+\dfrac{4}{14}$

7) $\dfrac{11}{19}+\dfrac{1}{19}$

8) $\dfrac{3}{16}+\dfrac{9}{16}$

9) $\dfrac{3}{10}+\dfrac{1}{10}$

10) $\dfrac{6}{17}+\dfrac{2}{17}$

11) $\dfrac{5}{22}+\dfrac{5}{22}$

12) $\dfrac{7}{35}+\dfrac{11}{35}$

13) $\dfrac{7}{27}+\dfrac{20}{27}$

14) $\dfrac{2}{31}+\dfrac{10}{31}$

15) $\dfrac{5}{23}+\dfrac{3}{23}$

16) $\dfrac{8}{41}+\dfrac{13}{41}$

17) $\dfrac{15}{37}+\dfrac{18}{37}$

18) $\dfrac{2}{51}+\dfrac{7}{51}$

19) $\dfrac{17}{26}+\dfrac{6}{26}$

20) $\dfrac{12}{48}+\dfrac{11}{48}$

21) $\dfrac{11}{29}+\dfrac{8}{29}$

22) $\dfrac{15}{34}+\dfrac{19}{34}$

23) $\dfrac{1}{19}+\dfrac{5}{19}$

24) $\dfrac{3}{53}+\dfrac{4}{53}$

25) $\dfrac{3}{20}+\dfrac{6}{20}$

26) $\dfrac{2}{63}+\dfrac{6}{63}$

27) $\dfrac{6}{38}+\dfrac{1}{38}$

28) $\dfrac{14}{31}+\dfrac{17}{31}$

29) $\dfrac{3}{28}+\dfrac{5}{28}$

30) $\dfrac{2}{37}+\dfrac{15}{37}$

FSA Subject Test Mathematics Grade 5

✏️ **Subtract fractions.**

1) $\dfrac{8}{9} - \dfrac{4}{9}$

2) $\dfrac{3}{8} - \dfrac{1}{8}$

3) $\dfrac{9}{11} - \dfrac{3}{11}$

4) $\dfrac{9}{14} - \dfrac{4}{14}$

5) $\dfrac{15}{20} - \dfrac{8}{20}$

6) $\dfrac{8}{15} - \dfrac{7}{15}$

7) $\dfrac{11}{19} - \dfrac{9}{19}$

8) $\dfrac{13}{16} - \dfrac{1}{16}$

9) $\dfrac{7}{29} - \dfrac{4}{29}$

10) $\dfrac{14}{23} - \dfrac{7}{23}$

11) $\dfrac{15}{34} - \dfrac{7}{34}$

12) $\dfrac{18}{41} - \dfrac{9}{41}$

13) $\dfrac{17}{39} - \dfrac{16}{39}$

14) $\dfrac{6}{26} - \dfrac{2}{26}$

15) $\dfrac{14}{17} - \dfrac{4}{17}$

16) $\dfrac{33}{55} - \dfrac{20}{55}$

17) $\dfrac{41}{49} - \dfrac{36}{49}$

18) $\dfrac{40}{53} - \dfrac{39}{53}$

19) $\dfrac{27}{37} - \dfrac{17}{37}$

20) $\dfrac{21}{47} - \dfrac{11}{47}$

21) $\dfrac{24}{43} - \dfrac{12}{43}$

22) $\dfrac{13}{19} - \dfrac{12}{19}$

23) $\dfrac{6}{26} - \dfrac{3}{26}$

24) $\dfrac{9}{15} - \dfrac{7}{15}$

25) $\dfrac{8}{39} - \dfrac{3}{39}$

26) $\dfrac{18}{61} - \dfrac{15}{61}$

27) $\dfrac{12}{53} - \dfrac{9}{53}$

28) $\dfrac{75}{76} - \dfrac{74}{76}$

29) $\dfrac{26}{45} - \dfrac{13}{45}$

30) $\dfrac{20}{57} - \dfrac{17}{57}$

Compare Fractions with Like Denominators

✏️ Evaluate and compare. Write < or > or =.

1) $\dfrac{1}{3} + \dfrac{1}{3} \underline{\quad} \dfrac{1}{3}$

2) $\dfrac{3}{6} + \dfrac{3}{6} \underline{\quad} \dfrac{5}{6}$

3) $\dfrac{8}{9} - \dfrac{4}{9} \underline{\quad} \dfrac{7}{9}$

4) $\dfrac{4}{11} + \dfrac{5}{11} \underline{\quad} \dfrac{7}{11}$

5) $\dfrac{9}{14} - \dfrac{8}{14} \underline{\quad} \dfrac{5}{14}$

6) $\dfrac{11}{17} - \dfrac{3}{17} \underline{\quad} \dfrac{6}{17}$

7) $\dfrac{11}{21} + \dfrac{2}{21} \underline{\quad} \dfrac{10}{21}$

8) $\dfrac{8}{32} + \dfrac{6}{32} \underline{\quad} \dfrac{9}{32}$

9) $\dfrac{25}{29} - \dfrac{16}{29} \underline{\quad} \dfrac{11}{29}$

10) $\dfrac{28}{41} + \dfrac{13}{41} \underline{\quad} \dfrac{27}{41}$

11) $\dfrac{18}{35} - \dfrac{11}{35} \underline{\quad} \dfrac{22}{35}$

12) $\dfrac{32}{47} - \dfrac{22}{47} \underline{\quad} \dfrac{11}{47}$

13) $\dfrac{14}{27} + \dfrac{13}{27} \underline{\quad} \dfrac{24}{27}$

14) $\dfrac{34}{52} - \dfrac{11}{52} \underline{\quad} \dfrac{21}{52}$

15) $\dfrac{43}{56} - \dfrac{24}{56} \underline{\quad} \dfrac{27}{56}$

16) $\dfrac{27}{71} + \dfrac{25}{71} \underline{\quad} \dfrac{48}{71}$

FSA Subject Test Mathematics Grade 5

More Than Two Fractions with Like Denominators

✎ Add fractions.

1) $\dfrac{5}{9} + \dfrac{2}{9} + \dfrac{2}{9}$

2) $\dfrac{4}{6} + \dfrac{1}{6} + \dfrac{1}{6}$

3) $\dfrac{2}{17} + \dfrac{4}{17} + \dfrac{2}{17}$

4) $\dfrac{1}{5} + \dfrac{1}{5} + \dfrac{1}{5}$

5) $\dfrac{7}{18} + \dfrac{2}{18} + \dfrac{3}{18}$

6) $\dfrac{3}{27} + \dfrac{5}{27} + \dfrac{2}{27}$

7) $\dfrac{4}{33} + \dfrac{4}{33} + \dfrac{4}{33}$

8) $\dfrac{8}{23} + \dfrac{6}{23} + \dfrac{2}{23}$

9) $\dfrac{13}{41} + \dfrac{2}{41} + \dfrac{8}{41}$

10) $\dfrac{6}{35} + \dfrac{9}{35} + \dfrac{20}{35}$

11) $\dfrac{1}{37} + \dfrac{5}{37} + \dfrac{5}{37}$

12) $\dfrac{4}{43} + \dfrac{9}{43} + \dfrac{8}{43}$

13) $\dfrac{4}{51} + \dfrac{10}{51} + \dfrac{7}{51}$

14) $\dfrac{5}{26} + \dfrac{13}{26} + \dfrac{6}{26}$

15) $\dfrac{5}{64} + \dfrac{4}{64} + \dfrac{2}{64}$

16) $\dfrac{1}{73} + \dfrac{5}{73} + \dfrac{6}{73}$

FSA Subject Test Mathematics Grade 5

Unlike Denominators

✏️ Add fraction.

1) $\dfrac{2}{9} + \dfrac{3}{4}$

2) $\dfrac{1}{4} + \dfrac{3}{5}$

3) $\dfrac{1}{16} + \dfrac{3}{4}$

4) $\dfrac{3}{8} + \dfrac{1}{7}$

5) $\dfrac{1}{3} + \dfrac{2}{4}$

6) $\dfrac{1}{6} + \dfrac{3}{7}$

7) $\dfrac{5}{18} + \dfrac{4}{6}$

8) $\dfrac{1}{12} + \dfrac{5}{6}$

9) $\dfrac{5}{27} + \dfrac{1}{9}$

10) $\dfrac{1}{6} + \dfrac{7}{24}$

11) $\dfrac{3}{5} + \dfrac{1}{8}$

12) $\dfrac{11}{42} + \dfrac{3}{7}$

13) $\dfrac{7}{20} + \dfrac{1}{3}$

14) $\dfrac{1}{45} + \dfrac{3}{5}$

15) $\dfrac{3}{32} + \dfrac{5}{8}$

16) $\dfrac{3}{48} + \dfrac{5}{6}$

17) $\dfrac{5}{12} + \dfrac{1}{6}$

18) $\dfrac{1}{34} + \dfrac{3}{17}$

19) $\dfrac{4}{9} + \dfrac{7}{54}$

20) $\dfrac{13}{56} + \dfrac{4}{7}$

21) $\dfrac{3}{12} + \dfrac{2}{3}$

22) $\dfrac{4}{33} + \dfrac{5}{11}$

WWW.MathNotion.Com

FSA Subject Test Mathematics Grade 5

✏️ **Subtract fractions.**

1) $\dfrac{8}{9} - \dfrac{1}{2}$

2) $\dfrac{2}{3} - \dfrac{3}{10}$

3) $\dfrac{1}{6} - \dfrac{1}{9}$

4) $\dfrac{7}{8} - \dfrac{1}{4}$

5) $\dfrac{3}{4} - \dfrac{1}{28}$

6) $\dfrac{11}{30} - \dfrac{3}{15}$

7) $\dfrac{11}{18} - \dfrac{5}{9}$

8) $\dfrac{5}{13} - \dfrac{3}{26}$

9) $\dfrac{17}{35} - \dfrac{2}{7}$

10) $\dfrac{5}{6} - \dfrac{12}{36}$

11) $\dfrac{5}{9} - \dfrac{1}{27}$

12) $\dfrac{3}{5} - \dfrac{1}{8}$

13) $\dfrac{2}{3} - \dfrac{3}{5}$

14) $\dfrac{7}{8} - \dfrac{3}{7}$

15) $\dfrac{5}{9} - \dfrac{13}{45}$

16) $\dfrac{3}{4} - \dfrac{5}{36}$

17) $\dfrac{39}{49} - \dfrac{5}{7}$

18) $\dfrac{3}{11} - \dfrac{3}{22}$

19) $\dfrac{17}{48} - \dfrac{4}{12}$

20) $\dfrac{2}{3} - \dfrac{4}{13}$

21) $\dfrac{5}{8} - \dfrac{19}{72}$

22) $\dfrac{3}{5} - \dfrac{1}{12}$

Ordering Fractions

✎ Order the fractions from least to greatest.

1) $\frac{1}{5}, \frac{1}{11}, \frac{1}{8}, \frac{1}{3}$ ____, ____, ____, ____

2) $\frac{1}{9}, \frac{1}{18}, \frac{2}{4}, \frac{1}{5}$ ____, ____, ____, ____

3) $\frac{4}{7}, \frac{1}{7}, \frac{6}{21}, \frac{15}{21}$ ____, ____, ____, ____

4) $\frac{1}{2}, \frac{1}{3}, \frac{4}{9}, \frac{5}{18}$ ____, ____, ____, ____

5) $\frac{4}{9}, \frac{3}{4}, \frac{7}{36}, \frac{1}{6}$ ____, ____, ____, ____

✎ Order the fractions from greatest to least.

6) $\frac{3}{4}, \frac{4}{7}, \frac{3}{10}, \frac{5}{13}$ ____, ____, ____, ____

7) $\frac{5}{11}, \frac{5}{6}, \frac{2}{5}, \frac{1}{3}$ ____, ____, ____, ____

8) $\frac{7}{8}, \frac{1}{6}, \frac{3}{4}, \frac{5}{15}$ ____, ____, ____, ____

9) $\frac{4}{7}, \frac{2}{3}, \frac{11}{25}, \frac{13}{33}$ ____, ____, ____, ____

10) $\frac{18}{20}, \frac{15}{16}, \frac{14}{18}, \frac{5}{12}$ ____, ____, ____, ____

FSA Subject Test Mathematics Grade 5

Denominators of 10, 100, and 1000

✎ Add fractions.

1) $\dfrac{7}{10}+\dfrac{13}{100}$

2) $\dfrac{1}{10}+\dfrac{10}{100}$

3) $\dfrac{15}{100}+\dfrac{1}{1,000}$

4) $\dfrac{56}{100}+\dfrac{3}{10}$

5) $\dfrac{50}{1,000}+\dfrac{7}{10}$

6) $\dfrac{6}{10}+\dfrac{30}{1,000}$

7) $\dfrac{9}{100}+\dfrac{3}{10}$

8) $\dfrac{5}{10}+\dfrac{50}{100}$

9) $\dfrac{48}{100}+\dfrac{6}{10}$

10) $\dfrac{70}{100}+\dfrac{2}{10}$

11) $\dfrac{80}{100}+\dfrac{200}{1,000}$

12) $\dfrac{30}{100}+\dfrac{4}{10}$

13) $\dfrac{9}{100}+\dfrac{7}{10}$

14) $\dfrac{25}{100}+\dfrac{6}{10}$

15) $\dfrac{15}{100}+\dfrac{8}{10}$

16) $\dfrac{3}{10}+\dfrac{31}{100}$

17) $\dfrac{8}{10}+\dfrac{11}{100}$

18) $\dfrac{34}{100}+\dfrac{6}{10}$

WWW.MathNotion.Com

FSA Subject Test Mathematics Grade 5

📝 Subtract fractions.

1) $\dfrac{8}{10} - \dfrac{20}{100}$

2) $\dfrac{5}{10} - \dfrac{47}{100}$

3) $\dfrac{12}{100} - \dfrac{60}{1,000}$

4) $\dfrac{6}{10} - \dfrac{50}{100}$

5) $\dfrac{3}{10} - \dfrac{23}{100}$

6) $\dfrac{70}{100} - \dfrac{250}{1,000}$

7) $\dfrac{4}{10} - \dfrac{350}{1,000}$

8) $\dfrac{70}{100} - \dfrac{3}{10}$

9) $\dfrac{40}{100} - \dfrac{3}{10}$

10) $\dfrac{6}{10} - \dfrac{180}{1,000}$

11) $\dfrac{93}{100} - \dfrac{5}{10}$

12) $\dfrac{65}{100} - \dfrac{4}{10}$

13) $\dfrac{80}{100} - \dfrac{6}{10}$

14) $\dfrac{90}{100} - \dfrac{5}{10}$

15) $\dfrac{200}{1,000} - \dfrac{1}{10}$

16) $\dfrac{90}{100} - \dfrac{7}{10}$

17) $\dfrac{900}{1,000} - \dfrac{40}{100}$

18) $\dfrac{60}{100} - \dfrac{3}{10}$

FSA Subject Test Mathematics Grade 5

Fractions to Mixed Numbers

✎ Convert fractions to mixed numbers.

1) $\dfrac{9}{5}$

2) $\dfrac{11}{3}$

3) $\dfrac{39}{8}$

4) $\dfrac{27}{11}$

5) $\dfrac{7}{2}$

6) $\dfrac{43}{4}$

7) $\dfrac{49}{9}$

8) $\dfrac{15}{4}$

9) $\dfrac{37}{7}$

10) $\dfrac{19}{7}$

11) $\dfrac{41}{9}$

12) $\dfrac{45}{12}$

13) $\dfrac{17}{5}$

14) $\dfrac{29}{6}$

15) $\dfrac{13}{4}$

16) $\dfrac{15}{7}$

17) $\dfrac{65}{7}$

18) $\dfrac{59}{8}$

19) $\dfrac{25}{4}$

20) $\dfrac{17}{8}$

WWW.MathNotion.Com

Mixed Numbers to Fractions

✎ Convert to fraction.

1) $3\frac{3}{5}$

2) $1\frac{1}{3}$

3) $4\frac{2}{5}$

4) $4\frac{2}{8}$

5) $2\frac{1}{5}$

6) $2\frac{8}{11}$

7) $4\frac{4}{7}$

8) $3\frac{7}{12}$

9) $2\frac{1}{3}$

10) $7\frac{5}{7}$

11) $2\frac{7}{10}$

12) $3\frac{4}{9}$

13) $1\frac{5}{8}$

14) $4\frac{3}{11}$

15) $3\frac{4}{7}$

16) $5\frac{2}{8}$

17) $7\frac{1}{7}$

18) $13\frac{1}{2}$

19) $4\frac{2}{7}$

20) $5\frac{2}{10}$

21) $12\frac{1}{3}$

22) $7\frac{1}{8}$

FSA Subject Test Mathematics Grade 5

Add and Subtract Mixed Numbers

✎ Add mixed numbers.

1) $3\frac{2}{5} + 8\frac{1}{5}$

2) $3\frac{2}{3} + 4\frac{1}{2}$

3) $6\frac{2}{7} + 2\frac{3}{7}$

4) $4\frac{2}{5} + 3\frac{1}{4}$

5) $8\frac{3}{4} - 2\frac{1}{2}$

6) $6\frac{5}{12} - 4\frac{1}{4}$

7) $5\frac{3}{8} - 3\frac{7}{8}$

8) $6\frac{1}{4} - 2\frac{15}{16}$

9) $9\frac{23}{28} - 4\frac{17}{28}$

10) $6\frac{1}{6} + 6\frac{2}{3}$

11) $4\frac{2}{9} + 5\frac{5}{9}$

12) $2\frac{1}{4} + 7\frac{4}{7}$

13) $7\frac{1}{5} - 3\frac{3}{5}$

14) $3\frac{1}{6} + 2\frac{3}{7}$

15) $2\frac{1}{3} + 4\frac{1}{4}$

16) $4\frac{1}{4} - 1\frac{2}{5}$

17) $\frac{1}{3} + 6\frac{1}{6}$

18) $2\frac{3}{5} + 2\frac{1}{10}$

WWW.MathNotion.Com

FSA Subject Test Mathematics Grade 5

Answers of Worksheets

Simplifying Fractions

1) $\frac{11}{21}$
2) $\frac{2}{5}$
3) $\frac{3}{4}$
4) $\frac{1}{6}$
5) $\frac{1}{2}$
6) $\frac{1}{7}$
7) $\frac{2}{7}$
8) $\frac{1}{3}$
9) $\frac{4}{5}$
10) $\frac{1}{7}$
11) $\frac{1}{3}$
12) $\frac{3}{2}$
13) $\frac{5}{7}$
14) $\frac{3}{4}$
15) $\frac{1}{5}$
16) $\frac{5}{8}$
17) $\frac{3}{8}$
18) $\frac{3}{8}$
19) $\frac{1}{9}$
20) $\frac{3}{4}$
21) $\frac{1}{9}$
22) $\frac{1}{2}$

Like Denominators (addition)

1) 1
2) 1
3) $\frac{11}{9}$
4) $\frac{4}{7}$
5) $\frac{7}{13}$
6) $\frac{5}{14}$
7) $\frac{12}{19}$
8) $\frac{3}{4}$
9) $\frac{2}{5}$
10) $\frac{8}{17}$
11) $\frac{5}{11}$
12) $\frac{18}{35}$
13) 1
14) $\frac{12}{31}$
15) $\frac{8}{23}$
16) $\frac{21}{41}$
17) $\frac{33}{37}$
18) $\frac{3}{17}$
19) $\frac{23}{26}$
20) $\frac{23}{48}$
21) $\frac{19}{29}$
22) 1
23) $\frac{6}{19}$
24) $\frac{7}{53}$
25) $\frac{9}{20}$
26) $\frac{8}{63}$
27) $\frac{7}{38}$
28) 1
29) $\frac{2}{7}$
30) $\frac{17}{37}$

Like Denominators (Subtraction)

1) $\frac{4}{9}$
2) $\frac{1}{4}$
3) $\frac{6}{11}$
4) $\frac{5}{14}$
5) $\frac{7}{20}$
6) $\frac{1}{15}$
7) $\frac{2}{19}$
8) $\frac{3}{4}$
9) $\frac{3}{29}$
10) $\frac{7}{23}$
11) $\frac{8}{34}$
12) $\frac{9}{41}$
13) $\frac{1}{39}$
14) $\frac{2}{13}$
15) $\frac{10}{17}$
16) $\frac{13}{55}$
17) $\frac{5}{49}$
18) $\frac{1}{53}$
19) $\frac{10}{37}$
20) $\frac{10}{47}$
21) $\frac{12}{43}$
22) $\frac{1}{19}$
23) $\frac{3}{26}$
24) $\frac{2}{15}$

WWw.MathNotion.Com

FSA Subject Test Mathematics Grade 5

25) $\frac{5}{39}$ 27) $\frac{3}{53}$ 29) $\frac{13}{45}$

26) $\frac{3}{61}$ 28) $\frac{1}{76}$ 30) $\frac{1}{19}$

Compare Fractions with Like Denominators

1) $\frac{2}{3} > \frac{1}{3}$ 5) $\frac{1}{14} < \frac{5}{14}$ 9) $\frac{9}{29} < \frac{11}{29}$ 13) $1 > \frac{24}{27}$

2) $1 > \frac{5}{6}$ 6) $\frac{8}{17} > \frac{6}{17}$ 10) $1 > \frac{27}{41}$ 14) $\frac{23}{52} > \frac{21}{52}$

3) $\frac{4}{9} < \frac{7}{9}$ 7) $\frac{13}{21} > \frac{10}{21}$ 11) $\frac{7}{35} < \frac{22}{35}$ 15) $\frac{19}{56} < \frac{27}{56}$

4) $\frac{9}{11} > \frac{7}{11}$ 8) $\frac{14}{32} > \frac{9}{32}$ 12) $\frac{10}{47} < \frac{11}{47}$ 16) $\frac{52}{71} > \frac{48}{71}$

More Than Two Fractions with Like Denominators

1) 1   5) $\frac{2}{3}$ 9) $\frac{23}{41}$ 13) $\frac{7}{17}$

2) 1   6) $\frac{10}{27}$ 10) 1   14) $\frac{12}{13}$

3) $\frac{8}{17}$ 7) $\frac{4}{11}$ 11) $\frac{11}{37}$ 15) $\frac{11}{64}$

4) $\frac{3}{5}$ 8) $\frac{16}{23}$ 12) $\frac{21}{43}$ 16) $\frac{12}{73}$

Unlike Denominators (Addition)

1) $\frac{35}{36}$ 7) $\frac{17}{18}$ 13) $\frac{41}{60}$ 19) $\frac{31}{54}$

2) $\frac{17}{20}$ 8) $\frac{11}{12}$ 14) $\frac{28}{45}$ 20) $\frac{45}{56}$

3) $\frac{13}{16}$ 9) $\frac{8}{27}$ 15) $\frac{23}{32}$ 21) $\frac{11}{12}$

4) $\frac{29}{56}$ 10) $\frac{11}{24}$ 16) $\frac{43}{48}$ 22) $\frac{19}{33}$

5) $\frac{5}{6}$ 11) $\frac{29}{40}$ 17) $\frac{7}{12}$

6) $\frac{25}{42}$ 12) $\frac{29}{42}$ 18) $\frac{7}{34}$

Unlike Denominators (Subtraction)

1) $\frac{7}{18}$ 5) $\frac{5}{7}$ 9) $\frac{1}{5}$ 13) $\frac{1}{15}$

2) $\frac{11}{30}$ 6) $\frac{1}{6}$ 10) $\frac{1}{2}$ 14) $\frac{25}{56}$

3) $\frac{1}{18}$ 7) $\frac{1}{18}$ 11) $\frac{14}{27}$ 15) $\frac{4}{15}$

4) $\frac{5}{8}$ 8) $\frac{7}{26}$ 12) $\frac{19}{40}$ 16) $\frac{11}{18}$

WWW.MathNotion.Com

FSA Subject Test Mathematics Grade 5

17) $\frac{4}{49}$ 19) $\frac{1}{48}$ 21) $\frac{13}{36}$

18) $\frac{3}{22}$ 20) $\frac{14}{39}$ 22) $\frac{31}{60}$

Ordering Fractions

1) $\frac{1}{11}, \frac{1}{8}, \frac{1}{5}, \frac{1}{3}$

2) $\frac{1}{18}, \frac{1}{9}, \frac{1}{5}, \frac{2}{4}$

3) $\frac{1}{7}, \frac{6}{21}, \frac{4}{7}, \frac{15}{21}$

4) $\frac{5}{18}, \frac{1}{3}, \frac{4}{9}, \frac{1}{2}$

5) $\frac{1}{6}, \frac{7}{36}, \frac{4}{9}, \frac{3}{4}$

6) $\frac{3}{4}, \frac{4}{7}, \frac{5}{13}, \frac{3}{10}$

7) $\frac{5}{6}, \frac{5}{11}, \frac{2}{5}, \frac{1}{3}$

8) $\frac{7}{8}, \frac{3}{4}, \frac{5}{15}, \frac{1}{6}$

9) $\frac{2}{3}, \frac{4}{7}, \frac{11}{25}, \frac{13}{33}$

10) $\frac{15}{16}, \frac{18}{20}, \frac{14}{18}, \frac{5}{12}$

Denominators of 10, 100, and 1000

1) $\frac{83}{100}$ 6) $\frac{63}{100}$ 11) 1   16) $\frac{61}{100}$

2) $\frac{1}{5}$ 7) $\frac{39}{100}$ 12) $\frac{7}{10}$ 17) $\frac{91}{100}$

3) $\frac{151}{1,000}$ 8) 1   13) $\frac{79}{100}$ 18) $\frac{47}{50}$

4) $\frac{43}{50}$ 9) $\frac{27}{25}$ 14) $\frac{17}{20}$

5) $\frac{3}{4}$ 10) $\frac{9}{10}$ 15) $\frac{19}{20}$

Denominators of 10, 100, and 1000 (Subtract)

1) $\frac{3}{5}$ 6) $\frac{9}{20}$ 11) $\frac{43}{100}$ 16) $\frac{1}{5}$

2) $\frac{3}{100}$ 7) $\frac{1}{20}$ 12) $\frac{1}{4}$ 17) $\frac{1}{2}$

3) $\frac{3}{50}$ 8) $\frac{2}{5}$ 13) $\frac{1}{5}$ 18) $\frac{3}{10}$

4) $\frac{1}{10}$ 9) $\frac{1}{10}$ 14) $\frac{2}{5}$

5) $\frac{7}{100}$ 10) $\frac{21}{50}$ 15) $\frac{1}{10}$

Fractions to Mixed Numbers

1) $1\frac{4}{5}$ 5) $3\frac{1}{2}$ 9) $5\frac{2}{7}$ 13) $3\frac{2}{5}$

2) $3\frac{2}{3}$ 6) $10\frac{3}{4}$ 10) $2\frac{5}{7}$ 14) $4\frac{5}{6}$

3) $4\frac{7}{8}$ 7) $5\frac{4}{9}$ 11) $4\frac{5}{9}$ 15) $3\frac{1}{4}$

4) $2\frac{5}{11}$ 8) $3\frac{3}{4}$ 12) $3\frac{9}{12}$ 16) $2\frac{1}{7}$

FSA Subject Test Mathematics Grade 5

17) $9\frac{2}{7}$ 18) $7\frac{3}{8}$ 19) $6\frac{1}{4}$ 20) $2\frac{1}{8}$

Mixed Numbers to Fractions

1) $\frac{18}{5}$ 7) $\frac{32}{7}$ 13) $\frac{13}{8}$ 19) $\frac{30}{7}$

2) $\frac{4}{3}$ 8) $\frac{43}{12}$ 14) $\frac{47}{11}$ 20) $\frac{52}{10}$

3) $\frac{22}{5}$ 9) $\frac{7}{3}$ 15) $\frac{25}{7}$ 21) $\frac{37}{3}$

4) $\frac{34}{8}$ 10) $\frac{54}{7}$ 16) $\frac{42}{8}$ 22) $\frac{57}{8}$

5) $\frac{11}{5}$ 11) $\frac{27}{10}$ 17) $\frac{50}{7}$

6) $\frac{30}{11}$ 12) $\frac{31}{9}$ 18) $\frac{27}{2}$

Add and Subtract Mixed Numbers

1) $11\frac{3}{5}$ 6) $2\frac{1}{6}$ 11) $9\frac{7}{9}$ 16) $2\frac{17}{20}$

2) $8\frac{1}{6}$ 7) $1\frac{1}{2}$ 12) $9\frac{23}{28}$ 17) $6\frac{1}{2}$

3) $8\frac{5}{7}$ 8) $3\frac{5}{16}$ 13) $3\frac{3}{5}$ 18) $4\frac{7}{10}$

4) $7\frac{13}{20}$ 9) $5\frac{3}{14}$ 14) $5\frac{25}{42}$

5) $6\frac{1}{4}$ 10) $12\frac{5}{6}$ 15) $6\frac{7}{12}$

Chapter 5 : Decimals

Topics that you'll learn in this chapter:

- ✓ Adding and Subtracting Decimals,
- ✓ Multiplying and Dividing Decimals,
- ✓ Round decimals,
- ✓ Comparing Decimals,

FSA Subject Test Mathematics Grade 5

Adding and Subtracting Decimals

🖎 Add and subtract decimals.

1) 24.19
   $-\ 15.42$
   _____

2) 42.23
   $+\ 25.42$
   _____

3) 72.54
   $+\ 11.28$
   _____

4) 57.45
   $-\ 24.75$
   _____

5) 43.57
   $+\ 54.85$
   _____

6) 86.68
   $-\ 54.12$
   _____

🖎 Solve.

7) _____ + 2.7 = 8.1

8) 6.4 + _____ = 12.8

9) 7.9 + _____ = 17

10) 4.6 + _____ = 15.3

11) _____ + 9.4 = 15

12) _____ + 8.24 = 13.54

🖎 Order each set of numbers from least to greatest.

1) 0.4, 0.67, 0.44, 0.73, 0.51   ___, ___, ___, ___, ___, ___

2) 3.9, 6.1, 4.28, 7.02, 4.65   ___, ___, ___, ___, ___, ___

3) 1.9, 1.04, 0.79, 0.72, 0.09   ___, ___, ___, ___, ___, ___

4) 2.6, 5.2, 1.9, 4.01, 1.99, 3.2   ___, ___, ___, ___, ___, ___

5) 4.2, 6.1, 3.8, 5.7, 2.1, 2.8   ___, ___, ___, ___, ___, ___

6) 0.56, 0.87, 0.14, 1.24, 3.1   ___, ___, ___, ___, ___, ___

WWW.MathNotion.Com

FSA Subject Test Mathematics Grade 5

Multiplying and Dividing Decimals

✎ Find each product.

1) 1.5
   × 2.1
   _____

4) 6.5
   × 0.99
   _____

7) 4.8
   × 9.1
   _____

2) 4.6
   × 3.4
   _____

5) 12.1
   × 5.2
   _____

8) 22.35
   × 20
   _____

3) 6.3
   × 2.5
   _____

6) 3.4
   × 8.9
   _____

9) 15.25
   × 3.6
   _____

✎ Find each quotient.

10) 3.5 ÷ 0.85

15) 21.5 ÷ 1,000

11) 15.35 ÷ 4.6

16) 4.1 ÷ 100

12) 32.42 ÷ 8.8

17) 9.7 ÷ 10

13) 9.2 ÷ 3.4

18) 6.55 ÷ 1.25

14) 0.84 ÷ 0.1

19) 18.48 ÷ 11.2

FSA Subject Test Mathematics Grade 5

Rounding Decimals

✎ Round each decimal number to the nearest place indicated.

1) 0.3<u>2</u>

2) 5.<u>0</u>1

3) 8.<u>8</u>24

4) 0.<u>4</u>78

5) <u>7</u>.32

6) 0.<u>2</u>9

7) 11.<u>3</u>1

8) <u>6</u>.223

9) 9.6<u>3</u>7

10) 5.<u>4</u>804

11) <u>7</u>.9

12) <u>5</u>.2439

13) 6.<u>4</u>92

14) 1.<u>6</u>2

15) 7<u>2</u>.85

16) 8<u>3</u>.67

17) 41.<u>6</u>8

18) 79<u>4</u>.741

19) 5<u>2</u>.2

20) 7<u>6</u>.93

21) <u>3</u>.219

22) 7<u>2</u>.09

23) 486.<u>4</u>91

24) 7.<u>0</u>8

FSA Subject Test Mathematics Grade 5

Comparing Decimals

✏️ Write the correct comparison symbol (>, < or =).

1) 0.35 ___ 1.8

2) 1.9 ___ 1.19

3) 8.6 ___ 8.6

4) 2.45 ___ 24.5

5) 7.56 ___ 0.756

6) 11.4 ___ 11.05

7) 7.4 ___ 0.7.4

8) 8.56 ___ 0.85

9) 7 ___ 0.7

10) 7.12 ___ 0.712

11) 12.3 ___ 12.5

12) 4.67 ___ 4.68

13) 2.57 ___ 2.75

14) 3.46 ___ 0.346

15) 6.87 ___ 6.78

16) 0.89 ___ 0.98

17) 1.57 ___ 0.157

18) 0.092 ___ 0.091

19) 24.3 ___ 24.3

20) 0.17 ___ 0.71

21) 0.46 ___ 0.64

22) 0.2 ___ 0.08

23) 0.10 ___ 0.1

24) 3.52 ___ 31.5

WWW.MathNotion.Com

FSA Subject Test Mathematics Grade 5

Answers of Worksheets

Adding and Subtracting Decimals

1) 8.77   4) 32.7   7) 5.4   10) 10.7
2) 67.65   5) 98.42   8) 6.4   11) 5.6
3) 83.82   6) 32.56   9) 9.1   12) 5.3

Order and Comparing Decimals

1) 0.4, 0.44, 0.51, 0.67, 0.73   4) 1.9, 1.99, 2.6, 3.2, 4.01, 5.2
2) 3.9, 4.28, 4.65, 6.1, 7.02   5) 2.1, 2.8, 3.8, 4.2, 5.7, 6.1
3) 0.09, 0.72, 0.79, 1.04, 1.9   6) 0.14, 0.56, 0.87, 1.24, 3.1

Multiplying and Dividing Decimals

1) 3.15   6) 30.26   11) 3.336…   16) 0.041
2) 15.64   7) 43.68   12) 3.684…   17) 0.97
3) 15.75   8) 447   13) 2.705…   18) 5.24
4) 6.435   9) 54.9   14) 8.4   19) 1.65
5) 62.92   10) 4.117…   15) 0.0215

Rounding Decimals

1) 0.3   7) 11.3   13) 6.5   19) 52
2) 5.0   8) 6   14) 1.6   20) 77
3) 8.8   9) 9.64   15) 73   21) 3
4) 0.5   10) 5.5   16) 84   22) 72
5) 7   11) 8   17) 41.7   23) 486.5
6) 0.3   12) 5   18) 795   24) 7.1

Comparing Decimals

1) 0.35 < 1.8   9) 7 > 0.7   17) 1.57 > 0.157
2) 1.9 > 1.19   10) 7.12 > 0.712   18) 0.092 > 0.091
3) 8.6 = 8.6   11) 12.3 < 12.5   19) 24.3 = 24.3
4) 2.45 < 24.5   12) 4.67 < 4.68   20) 0.17 < 0.71
5) 7.56 > 0.756   13) 2.57 < 2.75   21) 0.46 < 0.64
6) 11.4 > 11.05   14) 3.46 > 0.346   22) 0.2 > 0.08
7) 7.4 > 0.74   15) 6.87 > 6.78   23) 0.10 = 0.1
8) 8.56 > 0.85   16) 0.89 < 0.98   24) 3.52 < 31.5

WWW.MathNotion.Com

Chapter 6 : Ratios and Rates

Topics that you'll learn in this chapter:

- ✓ Simplifying Ratios,
- ✓ Writing Ratios,
- ✓ Create a Proportion,
- ✓ Proportional Ratios,
- ✓ Similar Figures,
- ✓ Word Problems,

FSA Subject Test Mathematics Grade 5

Simplifying Ratios

✏️ Reduce each ratio.

1) 14: 56
2) 12: 36
3) 5: 35
4) 56: 48
5) 12: 14
6) 81: 63
7) 60: 3
8) 15: 10

9) 25: 20
10) 14: 28
11) 60: 84
12) 11: 99
13) 30: 45
14) 18: 45
15) 90: 15
16) 1.5: 3

17) 8: 88
18) 13: 52
19) 3: 75
20) 2.2: 22
21) 11: 33
22) 18: 81
23) 68: 80
24) 50: 500

Writing Ratios

✏️ Express each ratio as a rate and unite rate.

1) 180 miles on 6 gallons of gas.

2) 99 dollars for 11 books.

3) 35 miles on 3.5 gallons of gas

4) 7.5 inches of snow in 1.5 hours

✏️ Express each ratio as a fraction in the simplest form.

5) 6 feet out of 60 feet

6) 14 cakes out of 49 cakes

7) 32 dimes t0 60 dimes

8) 18 dimes out of 63 coins

9) 13 cups to 91 cups

10) 28 gallons to 42 gallons

11) 35 miles out of 120 miles

12) 22 blue cars out of 55 cars

13) 6.9 pennies to 69 pennies

14) 14 beetles out of 70 insects

15) 18 dimes to 54 dimes

16) 40 red cars out of 160 cars

WWW.MathNotion.Com

FSA Subject Test Mathematics Grade 5

Create a Proportion

✏️ Create proportion from the given set of numbers.

1) 1, 20, 4, 5

2) 9, 135, 1, 15

3) 5, 15, 8, 24

4) 49, 7, 4, 28

5) 9, 1, 108, 12

6) 45, 2, 5, 18

7) 28, 7, 24, 6

8) 11, 3, 55, 15

9) 5, 45, 36, 4

10) 16, 128, 1, 8

11) 2.5, 10, 5, 20

12) 5, 12, 15, 36

Proportional Ratios

✏️ Solve each proportion.

1) $\dfrac{4}{8} = \dfrac{5}{d}$

2) $\dfrac{k}{6} = \dfrac{3}{9}$

3) $\dfrac{10}{8} = \dfrac{12}{x}$

4) $\dfrac{x}{15} = \dfrac{9}{5}$

5) $\dfrac{d}{11} = \dfrac{10}{1.1}$

6) $\dfrac{4.5}{6} = \dfrac{9}{x}$

7) $\dfrac{7}{15} = \dfrac{k}{60}$

8) $\dfrac{11}{1.5} = \dfrac{121}{d}$

9) $\dfrac{x}{0.7} = \dfrac{15}{2.8}$

10) $\dfrac{1.2}{4} = \dfrac{x}{2.5}$

11) $\dfrac{7.8}{x} = \dfrac{7.8}{2}$

12) $\dfrac{x}{3.4} = \dfrac{48}{16}$

13) $\dfrac{80}{20} = \dfrac{k}{60}$

14) $\dfrac{1.4}{5} = \dfrac{28}{d}$

15) $\dfrac{x}{7} = \dfrac{30}{15}$

16) $\dfrac{4}{1.6} = \dfrac{k}{1.6}$

17) $\dfrac{0.8}{1.2} = \dfrac{5.6}{d}$

18) $\dfrac{25}{x} = \dfrac{50}{4}$

19) $\dfrac{d}{9} = \dfrac{18}{27}$

20) $\dfrac{k}{12.6} = \dfrac{5}{12.6}$

21) $\dfrac{1.4}{7} = \dfrac{x}{10}$

22) $\dfrac{13}{5} = \dfrac{k}{15}$

23) $\dfrac{18}{21} = \dfrac{x}{7}$

24) $\dfrac{9}{99} = \dfrac{x}{22}$

WWW.MathNotion.Com

Similar Figures

🖊 Each pair of figures is similar. Find the missing side.

1)

2)

3)

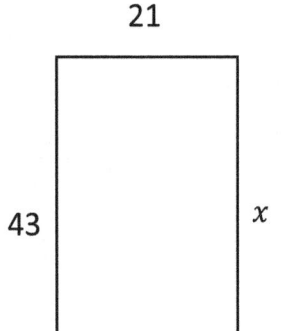

FSA Subject Test Mathematics Grade 5

Word Problems

✎ Solve.

1) In a party, 15 soft drinks are required for every 18 guests. If there are 360 guests, how many soft drinks is required?

2) In Jack's class, 16 of the students are tall and 10 are short. In Michael's class 40 students are tall and 25 students are short. Which class has a higher ratio of tall to short students?

3) Are these ratios equivalent?
12 cards to 84 animals 18 marbles to 126 marbles

4) The price of 6 apples at the Quick Market is $2.7. The price of 7 of the same apples at Walmart is $3.64. Which place is the better buy?

5) The bakers at a Bakery can make 160 bagels in 4 hours. How many bagels can they bake in 11 hours? What is that rate per hour?

FSA Subject Test Mathematics Grade 5

≫ Answer each question and round your answer to the nearest whole number.

6) If a 24.6 ft tall flagpole casts a 190.95 ft long shadow, then how long is the shadow that a 4.7 ft tall woman casts?

7) A model igloo has a scale of 2 in: 7 ft. If the real igloo is 56 ft wide, then how wide is the model igloo?

8) If an 88 ft tall tree casts a 8 ft long shadow, then how tall is an adult giraffe that casts a 4 ft shadow?

9) Find the distance between San Joe and Mount Pleasant if they are 5 cm apart on a map with a scale of 1 cm: 8 km.

10) A telephone booth that is 54 ft tall casts a shadow that is 9 ft long. Find the height of a lawn ornament that casts a 7 ft shadow.

FSA Subject Test Mathematics Grade 5

Answers of Worksheets

Simplifying Ratios

1) 2: 8	7) 20: 1	13) 2: 3	19) 1: 25
2) 1: 3	8) 3: 2	14) 2: 5	20) 1: 10
3) 1: 7	9) 5: 4	15) 6: 1	21) 1: 3
4) 7: 6	10) 1: 2	16) 1: 2	22) 2: 9
5) 6: 7	11) 5: 7	17) 1: 11	23) 17: 20
6) 9: 7	12) 1: 9	18) 1: 4	24) 1: 10

Writing Ratios

1) $\frac{180 \text{ miles}}{6 \text{ gallons}}$, 30 miles per gallon

2) $\frac{99 \text{ dollars}}{11 \text{ books}}$, 9.00 dollars per book

3) $\frac{35 \text{ miles}}{3.5 \text{ gallons}}$, 10 miles per gallon

4) $\frac{7.5" \text{ of snow}}{1.5 \text{ hours}}$, 5 inches of snow per hour

5) $\frac{1}{10}$	8) $\frac{2}{7}$	11) $\frac{7}{24}$	14) $\frac{1}{5}$
6) $\frac{2}{7}$	9) $\frac{1}{7}$	12) $\frac{2}{5}$	15) $\frac{1}{3}$
7) $\frac{8}{15}$	10) $\frac{2}{3}$	13) $\frac{1}{10}$	16) $\frac{1}{4}$

Create Proportion

1) 1: 5 = 4: 20	5) 9: 1 = 108: 12	9) 4: 5 = 36: 45
2) 9: 135 = 1: 15	6) 45: 18 = 5: 2	10) 128: 16 = 8: 1
3) 8: 5 = 24: 15	7) 24: 28 = 6: 7	11) 2.5: 5 = 10: 20
4) 7: 4 = 49: 28	8) 11: 3 = 55: 15	12) 15: 5 = 36: 12

Proportional Ratios

1) 10	7) 28	13) 240	19) 6
2) 2	8) 16.5	14) 100	20) 5
3) 9.6	9) 3.75	15) 14	21) 2
4) 27	10) 0.75	16) 4	22) 39
5) 100	11) 2	17) 8.4	23) 6
6) 12	12) 10.2	18) 2	24) 2

Similar Figures

1) 4	2) 2	3) 43

FSA Subject Test Mathematics Grade 5

Word Problems

1) 300

2) The ratio for both classes is equal to 8 to 5.

3) Yes! Both ratios are 1 to 7

4) The price at the Quick Market is a better buy.

5) 440, the rate is 40 per hour.

6) 36.48 ft

7) 16 in

8) 44 ft

9) 40 km

10) 42 ft

Chapter 7 : Measurement

Topics that you'll learn in this chapter:

- ✓ Reference Measurement Units,
- ✓ Metric Length Units,
- ✓ Customary Length Units,
- ✓ Metric Capacity Units,
- ✓ Customary Capacity Units,
- ✓ Metric Weight and Mass Units,
- ✓ Customary Weight and Mass Units,
- ✓ Temperature Units,
- ✓ Time,
- ✓ Add Money Amounts,
- ✓ Subtract Money Amounts,
- ✓ Money: Word Problems,

FSA Subject Test Mathematics Grade 5

Reference Measurement Units

LENGTH

Customary	Metric
1 mile (mi) = 1,760 yards (yd)	1 kilometer (km) = 1,000 meters (m)
1 yard (yd) = 3 feet (ft)	1 meter (m) = 100 centimeters (cm)
1 foot (ft) = 12 inches (in.)	1 centimeter(cm)= 10 millimeters(mm)

VOLUME AND CAPACITY

Customary	Metric
1 gallon (gal) = 4 quarts (qt)	1 liter (L) = 1,000 milliliters (mL)
1 quart (qt) = 2 pints (pt.)	
1 pint (pt.) = 2 cups (c)	
1 cup (c) = 8 fluid ounces (Fl oz)	

WEIGHT AND MASS

Customary	Metric
1 ton (T) = 2,000 pounds (lb.)	1 kilogram (kg) = 1,000 grams (g)
1 pound (lb.) = 16 ounces (oz)	1 gram (g) = 1,000 milligrams (mg)

Time

1 year = 12 months

1 year = 52 weeks

1 week = 7 days

1 day = 24 hours

1 hour = 60 minutes

1 minute = 60 seconds

FSA Subject Test Mathematics Grade 5

Metric Length Units

✏ Convert to the units.

1) 300 mm = _____ cm

2) 8 m = _____ mm

3) 4.5 m = _____ cm

4) 7 km = _____ m

5) 9,400mm = _____ m

6) 1,100 cm = _____ m

7) 2.8 m = _____ cm

8) 4,000 mm = _____ cm

9) 7,000 mm = _____ m

10) 2 km = _____ mm

11) 14.9 km = _____ m

12) 20 m = _____ cm

13) 5,000 m = _____ km

14) 7,600 m = _____ km

Customary Length Units

✏ Convert to the units.

1) 8 ft = _____ in

2) 4 ft = _____ in

3) 6 yd = _____ ft

4) 10 yd = _____ ft

5) 3,520 yd = _____ mi

6) 60 in = _____ ft

7) 144 in = ____ yd

8) 0.5 mi = _____ yd

9) 15 yd = _____ in

10) 42 yd = _____ in

11) 99 ft = _____ yd

12) 1.5 mi = _____ yd

13) 84 in = _____ ft

14) 30 yd = _____ feet

WWW.MathNotion.Com

FSA Subject Test Mathematics Grade 5

Metric Capacity Units

✎ Convert the following measurements.

1) 32.4 l = _____ ml

2) 7.1 l = _____ ml

3) 54 l = _____ ml

4) 92 l = _____ ml

5) 48 l = _____ ml

6) 13 l = _____ ml

7) 750 ml = _____ l

8) 2,400 ml = _____ l

9) 73,000 ml = _____ l

10) 8,000 ml = _____ l

11) 49,000 ml = _____ l

12) 5,500 ml = _____ l

Customary Capacity Units

✎ Convert the following measurements.

1) 51 gal = _____ qt.

2) 35 gal = _____ pt.

3) 68 gal = _____ c.

4) 20 pt. = _____ c

5) 12.5 qt = _____ pt.

6) 22.5 qt = _____ c

7) 51 pt. = _____ c

8) 48 c = _____ gal

9) 96 pt. = _____ gal

10) 136 qt = _____ gal

11) 15 c = _____ fl oz

12) 44 c = _____ qt

13) 240 c = _____ pt.

14) 148 qt = _____ gal

15) 160 pt. = _____ qt

16) 104 fl oz = _____ c.

WWW.MathNotion.Com

FSA Subject Test Mathematics Grade 5

Metric Weight and Mass Units

✏ Convert.

1) 60 kg = _____ g

2) 24 kg = _____ g

3) 610 kg = _____ g

4) 82 kg = _____ g

5) 95.8 kg = _____ g

6) 4.85 kg = _____ g

7) 1.5 kg = _____ g

8) 95,000 g = _____ kg

9) 241,000 g = _____ kg

10) 700,000 g = _____ kg

11) 2,500 g = _____ kg

12) 28,900 g = _____ kg

13) 970,000 g = _____ kg

14) 325,500 g = _____ kg

Customary Weight and Mass Units

✏ Convert.

1) 10,000 lb. = _____ T

2) 19,000 lb. = _____ T

3) 32,000 lb. = _____ T

4) 16,800 lb. = _____ T

5) 27 lb. = _____ oz

6) 25.4 lb. = _____ oz

7) 124 lb. = _____ oz

8) 4T = _____ lb.

9) 7T = _____ lb.

10) 11.2T = _____ lb.

11) 12.8T = _____ lb.

12) $\frac{6}{5}$ T = _____ oz

13) 9.125 T = _____ oz

14) $\frac{3}{4}$ T = _____ oz

WWW.MathNotion.Com

FSA Subject Test Mathematics Grade 5

Temperature Units

✎ Convert Fahrenheit into Celsius.

1) 21°F = ___ °C

2) 5.5°F = ___ °C

3) 71.6°F = ___ °C

4) 248°F = ___ °C

5) 75.2°F = ___ °C

6) 138.2°F = ___ °C

7) 167°F = ___ °C

8) 215.6°F = ___ °C

9) 104°F = ___ °C

10) 194°F = ___ °C

11) 203°F = ___ °C

12) 78.8°F = ___ °C

✎ Convert Celsius into Fahrenheit.

13) 2°C = ___ °F

14) 19°C = ___ °F

15) 48°C = ___ °F

16) 54°C = ___ °F

17) 81°C = ___ °F

18) 24°C = ___ °F

19) 88°C = ___ °F

20) 72°C = ___ °F

21) 80°C = ___ °F

22) 16°C = ___ °F

23) 10°C = ___ °F

24) 110°C = ___ °F

FSA Subject Test Mathematics Grade 5

Time

✏ Convert to the units.

1) 22 hr. = _____ min

2) 12.5 year = _____ week

3) 5.2hr = _____ sec

4) 21min = _____ sec

5) 1,800min = _____ hr.

6) 730 day = _____ year

7) 1.5year = _____ hr.

8) 42 day = _____ hr.

9) 2 day = _____ min

10) 660 min = _____ hr.

11) 10year = _____ month

12) 2,124sec = _____min

13) 168 hr = _____ day

14) 19 weeks = _____ day

✏ How much time has passed?

1) From 2:25 A.M. to 5:35 A.M.: ____ hours and ___ minutes.

2) From 3:30 A.M. to 7:55 A.M.: ____ hours and ___ minutes.

3) It's 6:30 P.M. What time was 2 hours ago? _____ O'clock

4) 4:30 A.M to 7:50 AM: _____ hours and _____ minutes.

5) 4:15 A.M to 8:35 AM: _____ hours and _____ minutes.

6) 5:10 A.M. to 7:35 AM. = _____ hour(s) and _____ minutes.

7) 10:55 A.M. to 3:25 PM. = _____ hour(s) and _____ minutes

8) 8:05 A.M. to 8:40 A.M. = _____ minutes

9) 6:02 A.M. to 6:49 A.M. = _____ minutes

WWW.MathNotion.Com

FSA Subject Test Mathematics Grade 5

Money Amounts

Add.

1) $128 + $328 $225 + $245 $150 + $186

2) $453 + $128 $440 + $541 $258 + $248

3) $645 + $112.5 $235.4 + $452.1 $125.99 + $148.32

4) $321.40 + $175.80 $458.10 + $752.65 $652.00 + $324.70

Subtract.

5) $725 − $334 $543 − $248 $349 − $122

6) $658.20 − $220.30 $752.10 − $452.15 $312.50 − $89.90

7) $315.90 − $220.10 $548.40 − $342.10 $968.40 − $324.50

8) Linda had $18.60. She bought some game tickets for $9.25. How much did she have left?

Money: Word Problems

✏️ Solve.

1) How many boxes of envelopes can you buy with $40 if one box costs $8?

2) After paying $7.22 for a salad, Ella has $45.86. How much money did she have before buying the salad?

3) How many packages of diapers can you buy with $96 if one package costs $6?

4) Last week James ran 28.5 miles more than Michael. James ran 59 miles. How many miles did Michael run?

5) Last Friday Jacob had $14.68. Over the weekend he received some money for cleaning the attic. He now has $38.95. How much money did he receive?

6) After paying $3.15 for a sandwich, Amelia has $48.69. How much money did she have before buying the sandwich?

FSA Subject Test Mathematics Grade 5

Answers of Worksheets

Metric length

1) 30 cm
2) 8,000 mm
3) 450 cm
4) 7,000 m
5) 9.4 m
6) 11 m
7) 280 cm
8) 400 cm
9) 7 m
10) 2,000,000 mm
11) 14,900 m
12) 2,000 cm
13) 5 km
14) 7.6 km

Customary Length

1) 96
2) 48
3) 18
4) 30
5) 2
6) 5
7) 4
8) 880
9) 540
10) 1,512
11) 33
12) 2,640
13) 7
14) 90

Metric Capacity

1) 32,400 ml
2) 7,100 ml
3) 54,000 ml
4) 92,000 ml
5) 48,000 ml
6) 13,000 ml
7) 0.75ml
8) 2.4 ml
9) 73 ml
10) 8L
11) 49 L
12) 5.5 L

Customary Capacity

1) 204 qt
2) 280 pt.
3) 1,088 c
4) 40 c
5) 25 pt.
6) 90c
7) 102 c
8) 3 gal
9) 12 gal
10) 34 gal
11) 120 qt
12) 11qt
13) 120 pt.
14) 37 gal
15) 80 qt
16) 13 pt.

Metric Weight and Mass

1) 60,000 g
2) 24,000 g
3) 610,000 g
4) 82,000 g
5) 95,800g
6) 4,850 g
7) 1,500 g
8) 95 kg
9) 241 kg
10) 700 kg
11) 2.5 kg
12) 28.9 kg
13) 970 kg
14) 325.5 kg

Customary Weight and Mass

1) 5 T
2) 9.5 T
3) 16 T
4) 8.4 T
5) 432 oz
6) 406.4 oz

FSA Subject Test Mathematics Grade 5

7) 1,984 oz
8) 8,000 lb.
9) 14,000 lb.
10) 22,400 lb.
11) 25,600 lb.
12) 38,400 oz
13) 292,000 oz
14) 24,000 oz

Temperature

1) −6.11°C
2) −14.72°C
3) 22°C
4) 120°C
5) 24°C
6) 59°C
7) 75°C
8) 102°C
9) 40°C
10) 90°C
11) 95°C
12) 26°C
13) 35.6°F
14) 66.2°F
15) 118.4°F
16) 129.2°F
17) 177.8°F
18) 75.2°F
19) 190.4°F
20) 161.6°F
21) 176°F
22) 60.8°F
23) 50°F
24) 230°F

Time - Convert

1) 1,320 min
2) 650 weeks
3) 18,720 sec
4) 1,260 sec
5) 30 hr
6) 2 year
7) 13,140 hr
8) 1,008 hr
9) 2,880 min
10) 11 hr
11) 120 months
12) 35.4 min
13) 7 days
14) 133 days

Time - Gap

1) 3:10
2) 4:25
3) 4:30 P.M.
4) 3:20
5) 4:20
6) 2:25
7) 4:30
8) 35 minutes
9) 47 minutes

Add Money

1) 456, 470, 336
2) 581, 981, 506
3) 757.5, 687.5, 274.31
4) 497.2, 1210.75, 976.7

Subtract Money

5) 391, 295, 227
6) 437.9, 299.95, 222.6
7) 95.8, 206.3, 643.9
8) $9.35

Money: word problem

1) 5
2) $53.08
3) 16
4) 30.5
5) 24.27
6) 51.84

WWW.MathNotion.Com

Chapter 8 : Algebraic Thinking

Topics that you'll learn in this chapter:

- ✓ Finding Rules,
- ✓ Algebraic Word Problems,
- ✓ Evaluating Expressions,
- ✓ Variables and Expressions,

Finding Rules

Complete the output.

1- **Rule:** the output is $x - 10.5$

Input	x	15	18	27	32.25	48.5
Output	y					

17) **Rule:** the output is $x \times 5\frac{1}{3}$

Input	x	3	9	15	21	33
Output	y					

2- **Rule:** the output is $x \div 9$

Input	x	513	387	342	198	126
Output	y					

Find a rule to write an expression.

3- **Rule:** _____

Input	x	4	14	19	24
Output	y	10	35	47.5	60

4- **Rule:** _____

Input	x	5	13	19.6	34.5
Output	y	14.4	22.4	29	43.9

5- **Rule:** _____

Input	x	72	96	132	230.4
Output	y	9	12	16.5	28.8

Algebraic Word Problems

Circle the number sentence that fits the problem. Then solve for x.

1) Mary had $42. Then she earned more money (x). Now she has $86.

   $42 + x = $86   OR $42 + $86 = x

   x = ____

2) Lisa had $35. Then she earned more money (x). Now she has $78.

   $35 + x = $78   OR $35 + $78 = x

   x = ____

3) Matthew had $37. Then he earned more money (x). Now he has $98.

   $37 + x = $98   OR $37 + $98 = x

   x = ____

4) Charlotte gave 19 of the cookies he had baked to a friend and now he has 45 cookies left.   45 − 19 = x   OR   x − 19 = 45

   x = ____

5) Mia gave 32 of the cookies she had baked to a friend and now she has 55 cookies left.   55 − 32 = x   OR   x − 32 = 55

   x = ____

6) Lucas gave 41 of the cookies he had baked to a friend and now he has 49 cookies left.   . 49 − 41 = x   OR   x − 41 = 49

   x = ____

FSA Subject Test Mathematics Grade 5

Evaluate Expressions

Simplify each algebraic expression.

1) $11 - x$, $x = 3$

2) $x + 14$, $x = 4$

3) $8 - 3x$, $x = 1$

4) $3x + \frac{1}{3}$, $x = \frac{1}{2}$

5) $2x + 18$, $x = 1.5$

6) $7 - 3x$, $x = 1.2$

7) $12 + 2x - 15$, $x = 3.5$

8) $25 - 5x$, $x = 2.2$

9) $\frac{44}{x} - 58$, $x = 0.4$

10) $\frac{x}{3} - 15 + x$, $x = 12.6$

11) $\frac{x}{7} + 9.5$, $x = 28.7$

12) $\frac{33}{x} - 5.2 + 2.1x$, $x = 3$

13) $2x - \frac{45}{x} - 12$, $x = 9$

14) $\frac{x}{17} - 1.8$, $x = 34$

15) $2(12.5x + 8)$, $x = 2$

16) $16x + 13x - 27 + 9$,

$x = 1.5$

17) $8.7 - \frac{16}{x} + 3x$,

$x = 4$

18) $5(3a - 2a)$,

$a = 5.5$

19) $14 - 2x + 16 - x$,

$x = 3.5$

20) $6x - 3 - x$,

$x = 1.6$

21) $18 - 2(2x + x)$, $x = 0.2$

WWW.MathNotion.Com

FSA Subject Test Mathematics Grade 5

Variables and Expressions

✎ Write a verbal expression for each algebraic expression.

1) $3a - 7b$

2) $8.2c^2 + 5d$

3) $x - 19.5$

4) $\frac{90}{6}$

5) $x^2 + y^3$

6) $4x + 9$

7) $x^2 - 6y + 15$

8) $x^3 + 7y^2 - 6$

9) $\frac{1}{5}x + \frac{1}{4}y - 11$

10) $\frac{1}{7}(x + 13) - 18.6y$

✎ Write an algebraic expression for each verbal expression.

11) 14 less than h

12) The product of 19 and a

13) The 23.2 divided by K

14) The product of 7 and the third power of x

15) 14 more than h to the sixth power

16) 30 more than triple d

17) One eighth the square of b

18) The difference of 42.5 and 3 times a number

19) 73 more than the cube of a number

20) one-quarters the cube of a number

FSA Subject Test Mathematics Grade 5

Answers of Worksheets

Finding Rules

1)
Input	x	15	18	27	32.25	48.5
Output	y	4.5	7.5	16.5	21.75	38

2)
Input	x	3	9	15	21	33
Output	y	16	48	80	112	176

3)
Input	x	513	387	342	198	126
Output	y	57	43	38	22	14

4) $y = 2.5x$ 5) $y = x + 9.4$ 6) $y = x \div 8$

Algebraic Word Problems

1) $\$42 + x = \$86$; $x = 44$ 4) $x - 19 = 45$; $x = 64$
2) $\$35 + x = \$78$; $x = 43$ 5) $x - 32 = 55$; $x = 87$
3) $\$37 + x = \$98$; $x = 61$ 6) $x - 41 = 49$; $x = 90$

Evaluating Expressions

1) 8   7) 4   13) 1   19) 19.5
2) 18   8) 14   14) 0.2   20) 5
3) 5   9) 52   15) 66   21) 16.8
4) $\frac{11}{6}$ 10) 1.8   16) 25.5
5) 21   11) 13.6   17) 16.7
6) 3.4   12) 12.1   18) 27.5

Variables and Expressions

1) 3 times a minus 7 times b.

2) 8.2 times c squared plus 5 times d.

3) a number minus 19.5.

4) the quotient of 90 and 6.

5) x squared plus y cubed.

6) the product of 4 and x plus 9.

WWW.MathNotion.Com

FSA Subject Test Mathematics Grade 5

7) x squared plus the product of 6 and y plus 15.

8) x cubed plus the product of 7 and y squared minus the product of 6 and y.

9) the sum of one–fifth of x and one–quarters of y, minus 11.

10) one–seventh of the sum of x and 13 minus the product of 18.6 and y.

11) $14 < h$

12) $19a$

13) $\dfrac{23.2}{K}$

14) $7x^3$

15) $14 > h^6$

16) $3d < 30$

17) $\dfrac{1}{8}b^2$

18) $42.5 - 3a$

19) $73 > a^3$

20) $\dfrac{1}{4}x^3$

FSA Subject Test Mathematics Grade 5

Chapter 9 : Geometric

Topics that you'll learn in this chapter:

- ✓ Identifying Angles,
- ✓ Estimate and Measure Angles,
- ✓ Polygon Names,
- ✓ Classify Triangles,
- ✓ Parallel Sides in Quadrilaterals,
- ✓ Identify Parallelograms,
- ✓ Identify Trapezoids,
- ✓ Identify Rectangles,
- ✓ Perimeter and Area of Squares,
- ✓ Perimeter and Area of rectangles,
- ✓ Area and Perimeter: Word Problems,
- ✓ Circumference, Diameter and Radius,
- ✓ Volume of Cubes and Rectangle Prisms,

Identifying Angles

✏️ Write the name of the angles (Acute, Right, Obtuse, and Straight).

1)

2)

3)

4)

5)

6)

7)

8)

FSA Subject Test Mathematics Grade 5

Estimate Angle Measurements

✏️ Estimate the approximate measurement of each angle in degrees.

1)

2)

3)

4)

5)

6)

7)

8)

WWW.MathNotion.Com

FSA Subject Test Mathematics Grade 5

Measure Angles with a Protractor

✎ Use protractor to measure the angles below.

1)

2)

3)

4)

✎ Use a protractor to draw angles for each measurement given.

1) 140°

2) 100°

3) 110°

4) 120°

5) 55°

Polygon Names

✎ Write name of polygons.

1)

2)

3)

4)

5)

6)

FSA Subject Test Mathematics Grade 5

Classify Triangles

✎ Classify the triangles by their sides and angles.

1)

2)

3)

4)

5)

6)

Parallel Sides in Quadrilaterals

✎ Write name of quadrilaterals.

1)

2)

3)

4)

5)

6)

FSA Subject Test Mathematics Grade 5

Identify Rectangles

✏️ Solve.

1) A rectangle has _____ sides and _____ angles.

2) Draw a rectangle that is 5.5 centimeters long and 2.5 centimeters wide. What is the perimeter?

3) Draw a rectangle 3.5 cm long and 1.5 cm wide.

4) Draw a rectangle whose length is 4.25 cm and whose width is 2.45 cm. What is the perimeter of the rectangle?

5) What is the perimeter of the rectangle?

7.2

5.8

FSA Subject Test Mathematics Grade 5

Perimeter: Find the Missing Side Lengths

 Find the missing side of each shape.

1) perimeter = 57.2

2) perimeter = 21.2

3) perimeter = 27.5

4) perimeter = 35.2

5) perimeter = 75.6

6) perimeter = 30.8

7) perimeter = 36.25

8) perimeter = 46.8

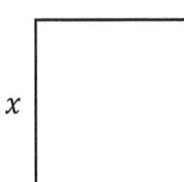

FSA Subject Test Mathematics Grade 5

Perimeter and Area of Squares

✏️ Find perimeter and area of squares.

1) A: _____, P: _____

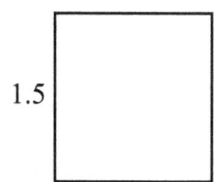

2) A: _____, P: _____

3) A: _____, P: _____

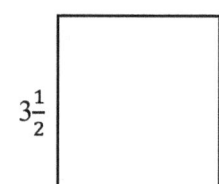

4) A: _____, P: _____

5) A: _____, P: _____

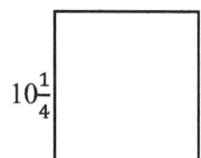

6) A: _____, P: _____

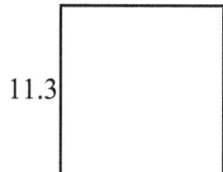

7) A: _____, P: _____

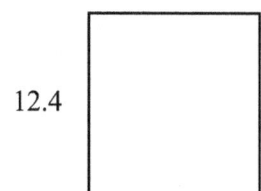

8) A: _____, P: _____

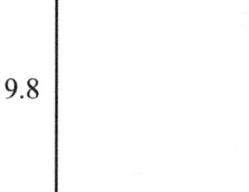

FSA Subject Test Mathematics Grade 5

Perimeter and Area of rectangles

🖎 Find perimeter and area of rectangles.

1) A: _____ , P: _____

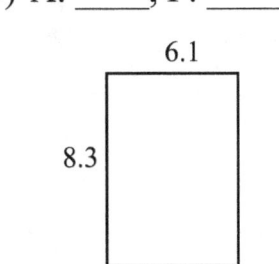

2) A: _____ , P: _____

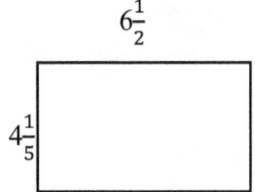

3) A: _____ , P: _____

6.1

8.3

4) A: _____ , P: _____

11.9

10.6

5) A: _____ , P: _____

6) A: _____ , P: _____

7) A: _____ , P: _____

8) A: _____ , P: _____

11.4

5.8

WWW.MathNotion.Com   99

FSA Subject Test Mathematics Grade 5

Find the Area or Missing Side Length of a Rectangle

✏ Find area or missing side length of rectangles.

1) Area =?

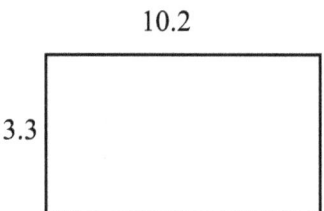

2) Area = 42.12, x=?

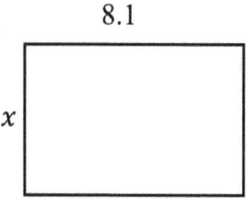

3) Area = 29.52, x=?

4) Area =?

5) Area =?

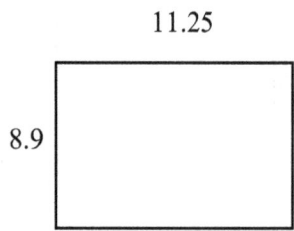

6) Area = 662.34 x=?

7) Area = 216.24, x=?

8) Area 336.42, x=?

WWW.MathNotion.Com

> **FSA Subject Test Mathematics Grade 5**

Area and Perimeter: Word Problems

✎ Solve.

1) The area of a rectangle is 90.86 square meters. The width is 7.7 meters. What is the length of the rectangle?

2) A square has an area of 6.25 square feet. What is the perimeter of the square?

3) Ava built a rectangular vegetable garden that is 3.2 feet long and has an area of 21.12 square feet. What is the perimeter of Ava's vegetable garden?

4) A square has a perimeter of 12.8 millimeters. What is the area of the square?

5) The perimeter of David's square backyard is 0.96 meters. What is the area of David's backyard?

6) The area of a rectangle is 37.63 square inches. The length is 7.1 inches. What is the perimeter of the rectangle?

Circumference, Diameter, and Radius

✎ Find the diameter and circumference of circles.

1)

2)

3)

4)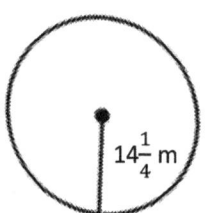

✎ Find the radius.

5)

6)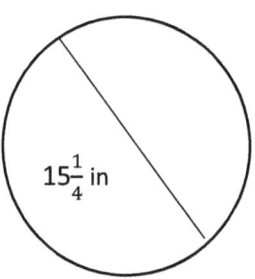

7) Diameter = $16\frac{1}{5} ft$
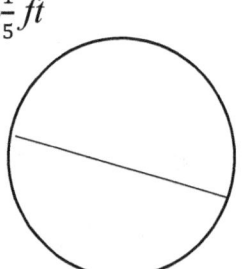

8) Diameter = 23.82 m
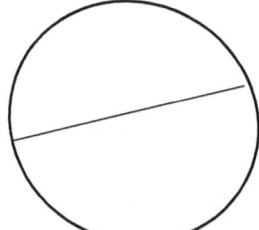

FSA Subject Test Mathematics Grade 5

Volume of Cubes and Rectangle Prisms

🖉 Find the volume of each of the rectangular prisms.

1)

2)

3)

4)

5)

6)

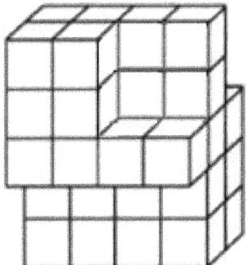

WWW.MathNotion.Com   103

FSA Subject Test Mathematics Grade 5

Answers of Worksheets

Identifying Angles

1) Right 3) Obtuse 5) Acute 7) Obtuse
2) Acute 4) Straight 6) Obtuse 8) Acute

Estimate Angle Measurements

1) 160° 3) 110° 5) 130° 7) 90°
2) 180° 4) 75° 6) 45° 8) 60°

Measure Angles with a Protractor

1) 50° 2) 135° 3) 20° 4) 170°

Draw Angles

1)   2)   3)

4)   5)

   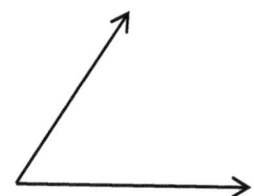

Polygon Names

1) Diamond 3) Pentagon 5) Heptagon
2) Parallelogram 4) Trapezius 6) Octagon

Classify Triangles

1) Scalene, acute   4) Scalene, right
2) Isosceles, acute   5) Isosceles, right
3) Equilateral, acute   6) Scalene, obtuse

WWW.MathNotion.Com

FSA Subject Test Mathematics Grade 5

Parallel Sides in Quadrilaterals

1) Hexagon
2) Kike
3) Parallelogram
4) Trapezoid
5) Rhombus
6) Rectangle

Identify Rectangles

1) 4 - 4
2) 16
3) Draw the rectangle.
4) 13.4
5) 26

Perimeter: Find the Missing Side Lengths

1) 14.3
2) 7.2
3) 5.5
4) 8.8
5) 18.9
6) 6.2
7) 7
8) 11.7

Perimeter and Area of Squares

1) A: 2.25, P: 6
2) A: 27.56, P: 21
3) A: 12.25, P: 14
4) A: 17.64, P: 16.8
5) A: 105.063 P: 41
6) A: 127.69, P: 45.2
7) A: 153.76, P: 49.6
8) A: 96.04, P: 39.2

Perimeter and Area of rectangles

1) A: 12.375, P: 15.5
2) A: 27.3, P: 21.4
3) A: 50.63, P: 28.8
4) A: 126.14, P: 45
5) A: 39.9, P: 25.7
6) A: 26.46, P: 21
7) A: 33.95, P: 26.4
8) A: 66.12, P: 34.4

Find the Area or Missing Side Length of a Rectangle

1) 33.66
2) 5.2
3) 7.2
4) 60.125
5) 100.125
6) 28.5
7) 10.2
8) 17.8

Area and Perimeter: Word Problems

1) 11.8
2) 10
3) 19.6
4) 10.24
5) 0.0576
6) 24.8

Circumference, Diameter, and Radius

1) diameter: 8.4 circumferences: 8.4π or 26.376
2) diameter: 23 circumferences: 23π or 72.22
3) diameter: 7.4 circumferences: 7.4π or 23.24
4) diameter: 28.5 circumferences: 28.5π or 89.49
5) radius: 0.009 in
6) radius: 7.625 in
7) radius: 8.1 ft
8) radius: 11.91 m

Volume of Cubes and Rectangle Prisms

1) 452.025 cm³
2) 429.2 cm³
3) 148.877 c m³
4) 583.848 cm³
5) 32
6) 40

WWW.MathNotion.Com

FSA Subject Test Mathematics Grade 5

Chapter 10 : Three-Dimensional Figures

Topics that you'll learn in this chapter:

- ✓ Identify Three–Dimensional Figures,
- ✓ Count Vertices, Edges, and Faces,
- ✓ Identify Faces of Three–Dimensional Figures,

Identify Three-Dimensional Figures

✎ Write the name of each shape.

1)

2)

3)

4)

5)

6)
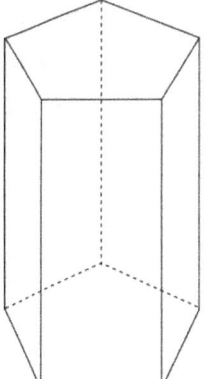

Count Vertices, Edges, and Faces

	Shape	Number of edges	Number of faces	Number of vertices
1)		_____	_____	_____
2)		_____	_____	_____
3)		_____	_____	_____
4)		_____	_____	_____
5)		_____	_____	_____
6)		_____	_____	_____

Identify Faces of Three-Dimensional Figures

✏️ Write the number of faces.

1)

2)

3)

4)

5)

6)

7)

8)

FSA Subject Test Mathematics Grade 5

Answers of Worksheets

Identify Three–Dimensional Figures

1) Cube
2) Triangular pyramid
3) Triangular prism
4) Square pyramid
5) Rectangular prism
6) Pentagonal prism
7) Hexagonal prism

Count Vertices, Edges, and Faces

Shape	Number of edges	Number of faces	Number of vertices
1)	6	4	4
2)	8	5	5
3)	12	6	8
4)	15	7	10
5)	12	6	8
6)	18	8	12

Identify Faces of Three–Dimensional Figures

1) 6
2) 2
3) 5
4) 4
5) 6
6) 7
7) 8
8) 5

FSA Subject Test Mathematics Grade 5

Chapter 11 : Symmetry and Transformations

Topics that you'll learn in this chapter:

- ✓ Line Segments,
- ✓ Identify Lines of Symmetry,
- ✓ Count Lines of Symmetry,
- ✓ Parallel, Perpendicular and Intersecting Lines,

Line Segments

✎ Write each as a line, ray, or line segment.

1)

2)

3)

4)

5)

6)

7)

8)

FSA Subject Test Mathematics Grade 5

Identify Lines of Symmetry

✎ Tell whether the line on each shape a line of symmetry is.

1)

2)

3)

4)

5)

6)

7)

8)

WWW.MathNotion.Com   113

FSA Subject Test Mathematics Grade 5

Count Lines of Symmetry

✍ Draw lines of symmetry on each shape. Count and write the lines of symmetry you see.

1)

2)

3)

4)

5)

6)

7)

8)

Parallel, Perpendicular and Intersecting Lines

✏ State whether the given pair of lines are parallel, perpendicular, or intersecting.

1)

2)

3)

4)

5)

6)

7)

8)
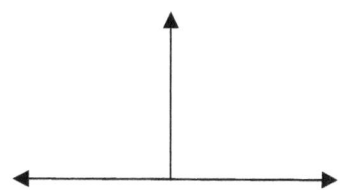

FSA Subject Test Mathematics Grade 5

Answers of Worksheets

Line Segments

1) Ray   4) Ray   7) Line
2) Line segment   5) Ray   8) Line segment
3) Line   6) Line

Identify lines of symmetry

1) yes   3) no   5) yes   7) no
2) no   4) yes   6) yes   8) yes

Count lines of symmetry

1)   2)   3)   4)

5)   6)   7)   8)

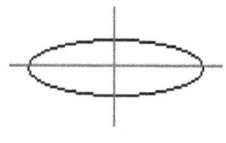

Parallel, Perpendicular and Intersecting Lines

1) Intersection   5) Intersection
2) Perpendicular   6) Perpendicular
3) Parallel   7) Parallel
4) Intersection   8) Perpendicular

WWW.MathNotion.Com

FSA Subject Test Mathematics Grade 5

Chapter 12 : Data Graphs, and Statistics

Topics that you'll learn in this chapter:

- ✓ Mean, Median, Mode, and Range,
- ✓ Graph Points on a Coordinate Plane,
- ✓ Bar Graph,
- ✓ Tally and Pictographs,
- ✓ Dot Plots
- ✓ Line Graphs,
- ✓ Stem–And–Leaf Plot,
- ✓ Scatter Plots,
- ✓ Probability Problems,

FSA Subject Test Mathematics Grade 5

Mean and Median

✎ Find Mean and Median of the Given Data.

1) 13, 22, 12, 6, 14

2) 7, 18, 11, 15, 2, 23

3) 33, 25, 14, 7, 14

4) 6, 7, 5, 1, 2, 5

5) 11, 4, 15, 7, 8, 17, 22

6) 9, 2, 5, 2, 18, 14, 45

7) 19, 14, 20, 12, 34, 14, 17, 31

8) 38, 29, 5, 3, 14, 9, 32

9) 28, 27, 31, 36, 32, 53, 41

10) 12, 17, 6, 17, 2, 17, 9, 12

11) 35, 11, 25, 54, 42, 41

12) 42, 48, 50, 48, 27, 67

13) 75, 72, 30, 46, 38, 29

14) 89, 73, 67, 46, 54, 84, 34

15) 78, 21, 100, 85, 54, 60

16) 41, 65, 9, 88, 17, 38, 14

✎ Solve.

17) In a javelin throw competition, five athletics score 54, 72, 59, 67 and 85 meters. What are their Mean and Median? _____

18) Eva went to shop and bought 18 apples, 7 peaches, 6 bananas, 3 pineapple and 9 melons. What are the Mean and Median of her purchase?

_____

FSA Subject Test Mathematics Grade 5

Mode and Range

✎ Find Mode and Rage of the Given Data.

1) 12, 4, 8, 2, 9, 4
   Mode: _____ Range: _____

2) 8, 8, 11, 8, 18, 2, 5, 41
   Mode: _____ Range: _____

3) 3, 3, 2, 23, 4, 12, 3, 9, 3
   Mode: _____ Range: _____

4) 12, 39, 5, 25, 4, 4, 27, 8
   Mode: _____ Range: _____

5) 5, 5, 8, 5, 14, 3, 12
   Mode: _____ Range: _____

6) 2, 5, 19, 15, 12, 11, 7, 8, 7, 7
   Mode: _____ Range: _____

7) 9, 4, 0, 12, 19, 21, 9, 7, 9, 3
   Mode: _____ Range: _____

8) 11, 4, 3, 11, 5, 8, 44, 11, 7
   Mode: _____ Range: _____

9) 3, 3, 6, 9, 3, 3, 9, 15, 14, 20
   Mode: _____ Range: _____

10) 15, 14, 14, 16, 20, 7, 1, 30
    Mode: _____ Range: _____

11) 13, 3, 27, 4, 4, 16, 18, 4
    Mode: _____ Range: _____

12) 7, 22, 35, 11, 7, 24, 6, 13
    Mode: _____ Range: _____

✎ Solve.

13) A stationery sold 14 pencils, 28 red pens, 51 blue pens, 14 notebooks, 27 erasers, 43 rulers and 40 color pencils. What are the Mode and Range for the stationery sells?

   Mode: _____   Range: _____

14) In an English test, eight students score 14, 20, 22, 14, 17, 28, 36 and 24. What are their Mode and Range? _____

WWW.MathNotion.Com

FSA Subject Test Mathematics Grade 5

Graph Points on a Coordinate Plane

Plot each point on the coordinate grid.

1) A (4, 6)   3) C (1, 5)   5) E (4, 8)
2) B (3, 2)   4) D (5, 7)   6) F (9, 2)

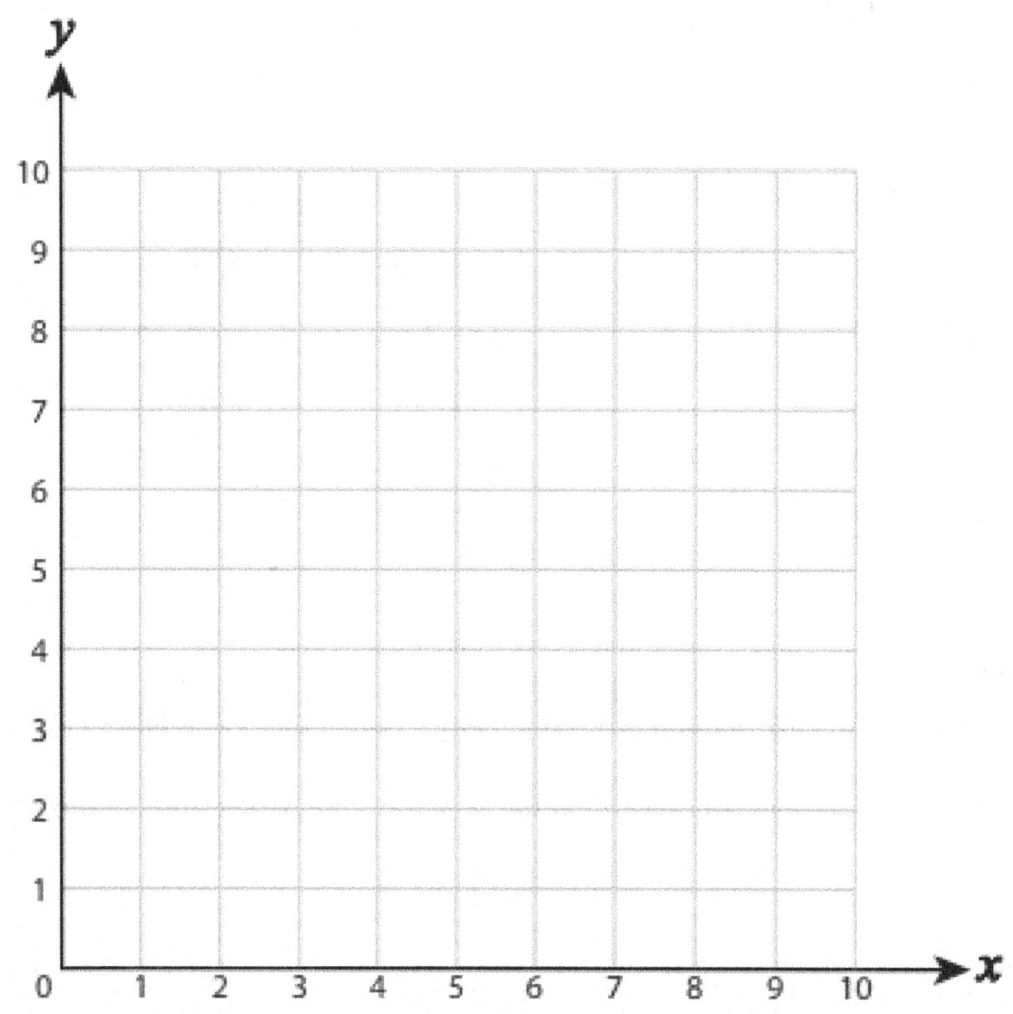

WWW.MathNotion.Com

Bar Graph

1) Graph the given information as a bar graph.

Day	Hot dogs sold
Monday	50
Tuesday	80
Wednesday	10
Thursday	30
Friday	70

Tally and Pictographs

✏️ Using the key, draw the pictograph to show the information.

Key: ⚽ = 2 animals

Dot plots

The ages of students in a Math class are given below.

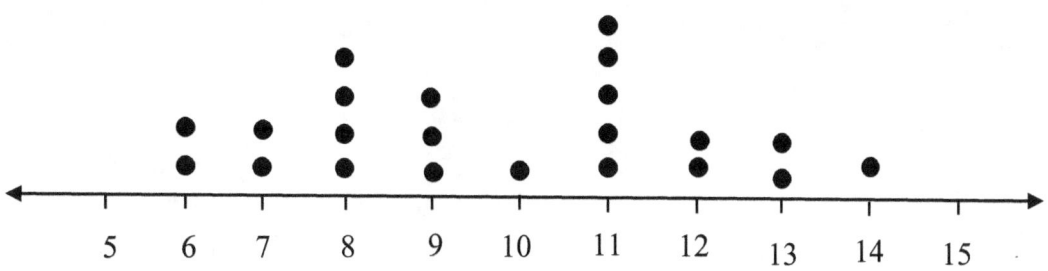

1) What is the total number of students in math class?

2) How many students are at least 12 years old?

3) Which age(s) has the most students?

4) Which age(s) has the fewest student?

5) Determine the median of the data.

6) Determine the range of the data.

7) Determine the mode of the data.

Line Graphs

David work as a salesman in a store. He records the number of shoes sold in five days on a line graph. Use the graph to answer the question

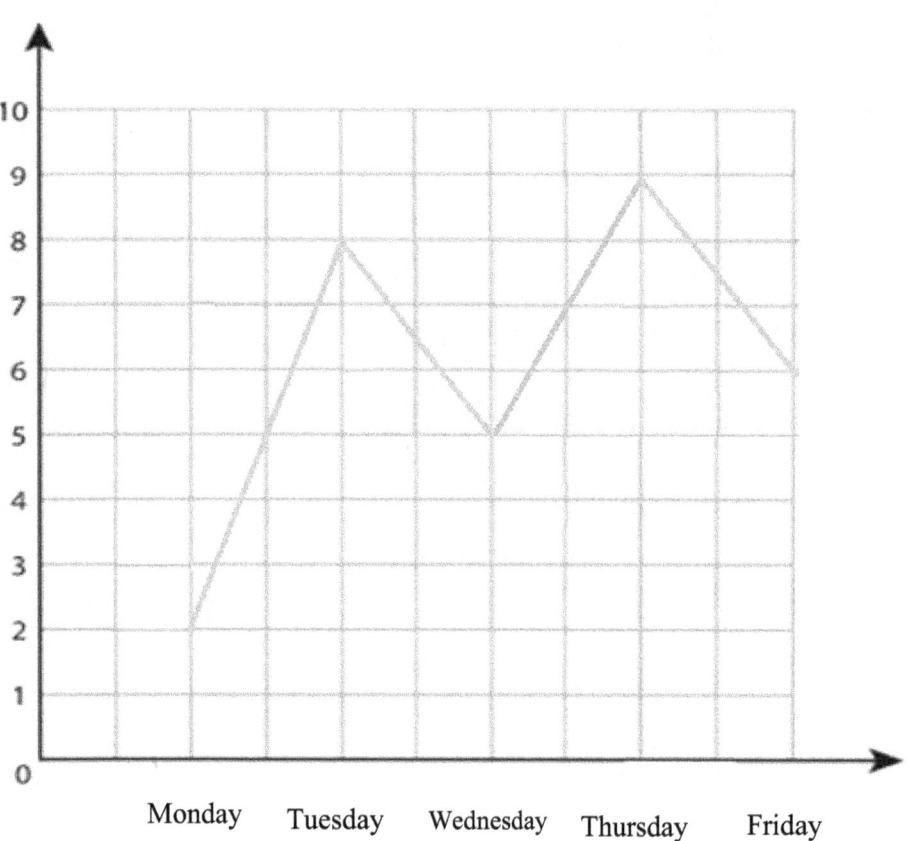

1) How many shoes were sold on Friday?

2) Which day had the minimum sales of shoes?

3) Which day had the maximum number of shoes sold?

4) How many shoes were sold in 5 days?

FSA Subject Test Mathematics Grade 5

Stem–And–Leaf Plot

✏️ Make stem ad leaf plots for the given data.

1) 42, 47, 14, 19, 42, 69, 65, 49, 42, 10, 64

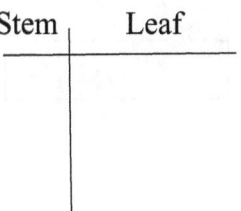

2) 43, 85, 52, 48, 45, 43, 51, 81, 59, 50, 85, 89

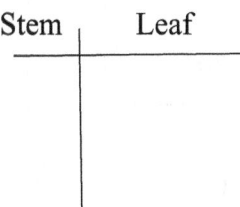

3) 112, 39, 46, 35, 80, 119, 42, 114, 37, 112, 47, 119

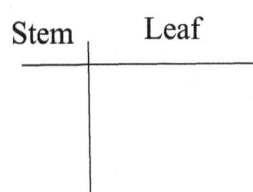

4) 90, 50, 131, 93, 112, 56, 139, 98, 115, 59, 98, 135, 111

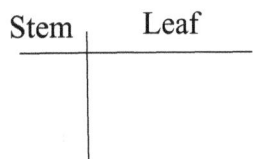

Scatter Plots

🖊 Construct a scatter plot.

x	1	2	3	4	5	8
y	15	40	50	35	70	20

Probability Problems

✏️ Solve.

1) A number is chosen at random from 1 to 20. Find the probability of selecting a 10 or smaller.

2) A number is chosen at random from 1 to 25. Find the probability of selecting multiples of 5.

3) A number is chosen at random from 1 to 15. Find the probability of selecting multiples of 7.

4) A number is chosen at random from 1 to 20. Find the probability of selecting a multiple of 6.

5) A number is chosen at random from 1 to 10. Find the probability of selecting prime numbers.

6) A number is chosen at random from 1 to 20. Find the probability of not selecting factors of 15.

FSA Subject Test Mathematics Grade 5

Answers of Worksheets

Mean and Median

1) Mean: 13.4, Median: 13
2) Mean: 12.67, Median: 13
3) Mean: 18.6, Median: 14
4) Mean: 4.33, Median: 5
5) Mean: 12, Median: 11
6) Mean: 13.57, Median: 9
7) Mean: 20.125, Median: 18
8) Mean: 18.57, Median: 14
9) Mean: 35.43, Median: 32

10) Mean: 11.5, Median: 12
11) Mean: 34.67, Median: 38
12) Mean: 47, Median: 48
13) Mean: 48.33, Median: 42
14) Mean: 63.86, Median: 67
15) Mean: 66.33, Median: 69
16) Mean: 38.86, Median: 38
17) Mean: 67.4, Median: 67
18) Mean: 8.6, Median: 7

Mode and Range

1) Mode: 4, Range: 10
2) Mode: 8, Range: 39
3) Mode: 3, Range: 21
4) Mode: 4, Range: 35
5) Mode: 5, Range: 11
6) Mode: 7, Range: 17
7) Mode: 9, Range: 21

8) Mode: 11, Range: 41
9) Mode: 3, Range: 17
10) Mode: 14, Range: 29
11) Mode: 4, Range: 24
12) Mode: 7, Range: 29
13) Mode: 14, Range: 37
14) Mode: 14, Range: 22

Graph Points on a Coordinate Plane

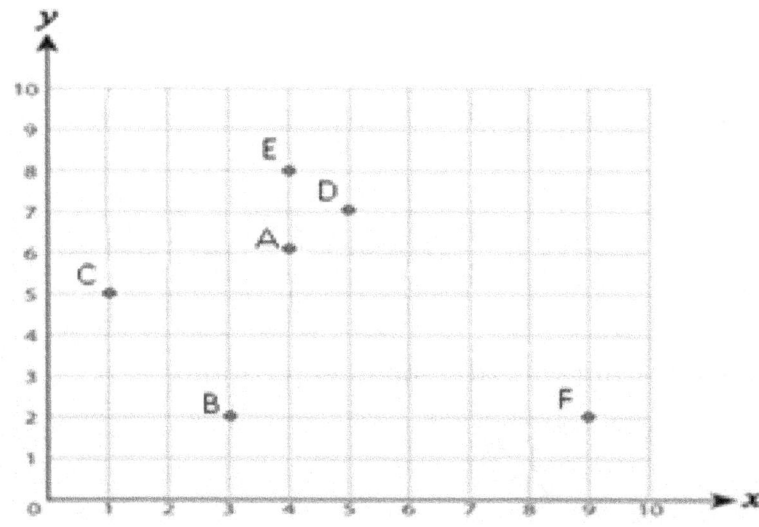

WWW.MathNotion.Com

FSA Subject Test Mathematics Grade 5

Bar Graph

Tally and Pictographs

Dot plots

1) 22
2) 5
3) 11
4) 10 and 14
5) 2
6) 4
7) 2

Line Graphs

1) 6
2) Monday
3) Thursday
4) 30

FSA Subject Test Mathematics Grade 5

Stem–And–Leaf Plot

1)
Stem	leaf
1	0 4 9
4	2 2 2 7 9
6	4 5 9

2)
Stem	leaf
4	3 3 5 8
5	0 1 2 9
8	1 5 5 9

3)
Stem	leaf
3	5 7 9
4	2 6 7
8	0
11	2 2 4 9 9

4)
Stem	leaf
5	0 6 9
9	0 3 8 8
11	1 2 5
13	1 5 9

Scatter Plots

Probability Problems

1) $\frac{1}{2}$

2) $\frac{1}{5}$

3) $\frac{2}{15}$

4) $\frac{3}{20}$

5) $\frac{2}{5}$

6) $\frac{4}{5}$

FSA Subject Test Mathematics Grade 5

Chapter 13 : FSA Math Practice Tests

Time to Test

Time to refine your skill with a practice examination.

Take a REAL FSA Mathematics test to simulate the test day experience. After you've finished, score your test using the answer key.

Before You Start

- You'll need a pencil and scratch papers to take the test.
- For this practice test, don't time yourself. Spend time as much as you need.
- It's okay to guess. You won't lose any points if you're wrong.
- After you've finished the test, review the answer key to see where you went wrong.

Calculators are not permitted for Grade 5 FSA Tests

Good Luck!

FSA Subject Test Mathematics Grade 5

FSA GRADE 5 MAHEMATICS REFRENCE MATERIALS

Perimeter

Square $\qquad$ $P = 4S$

Rectangle $\qquad$ $P = 2L + 2W$

Area

Square $\qquad$ $A = S \times S$

Rectangle $\qquad$ $A = l \times w$ or $A = bh$

Volume

Square $\qquad$ $A = S \times S \times S$

Rectangle $\qquad$ $A = l \times w \times h$ or $A = Bh$

LENGTH

Customary	Metric
1 mile (mi) = 1,760 yards (yd)	1 kilometer (km) = 1,000 meters (m)
1 yard (yd) = 3 feet (ft)	1 meter (m) = 100 centimeters (cm)
1 foot (ft) = 12 inches (in.)	1 centimeter (cm) = 10 millimeters (mm)

VOLUME AND CAPACITY

Customary	Metric
1 gallon (gal) = 4 quarts (qt)	1 liter (L) = 1,000 milliliters (mL)
1 quart (qt) = 2 pints (pt.)	
1 pint (pt.) = 2 cups (c)	
1 cup (c) = 8 fluid ounces (Fl oz)	

WEIGHT AND MASS

Customary	Metric
1 ton (T) = 2,000 pounds (lb.)	1 kilogram (kg) = 1,000 grams (g)
1 pound (lb.) = 16 ounces (oz)	1 gram (g) = 1,000 milligrams (mg)

WWW.MathNotion.Com

The Florida Standards Assessments

FSA Practice Test 1

Mathematics

GRADE 5

Released *Month Year*

Session 1

❖ **Calculators are NOT permitted for this practice test.**

❖ **Time for Session 1: 80 Minutes**

FSA Subject Test Mathematics Grade 5

1) The drivers at G & G trucking must report the mileage on their trucks each week. The mileage reading of Ed's vehicle was 49,895 at the beginning of one week, and 50,993 at the end of the same week. What was the total number of miles driven by Ed that week?

   A. 488 Miles

   B. 959 Miles

   C. 1,098 Miles

   D. 1,980 Miles

2) Camille uses a 25% off coupon when buying a sweater that costs $60. How much does she pay?

   A. $45

   B. $55

   C. $15

   D. $38

3) The area of a rectangle is D square feet and its length is 33 feet. Which equation represents W, the width of the rectangle in feet?

   A. $W = \dfrac{D}{33}$

   B. $W = \dfrac{33}{D}$

   C. $W = 33D$

   D. $W = 33 + D$

FSA Subject Test Mathematics Grade 5

4) A baker uses 6 eggs to bake a cake. How many cakes will he be able to bake with 540 eggs?

   A. 80

   B. 90

   C. 120

   D. 60

5) Which list shows the fractions in order from least to greatest?

$$\frac{5}{8}, \frac{8}{10}, \frac{1}{10}, \frac{5}{9}, \frac{3}{7}$$

   A. $\frac{5}{9}, \frac{3}{7}, \frac{8}{10}, \frac{1}{10}, \frac{5}{8}$

   B. $\frac{5}{8}, \frac{1}{10}, \frac{5}{9}, \frac{3}{7}, \frac{8}{10}$

   C. $\frac{8}{10}, \frac{5}{9}, \frac{3}{7}, \frac{1}{10}, \frac{5}{8}$

   D. $\frac{1}{10}, \frac{3}{7}, \frac{5}{9}, \frac{5}{8}, \frac{8}{10}$

6) Which statement about 6 multiplied by $\frac{3}{8}$ is true?

   A. The product is between 1 and 2.

   B. The product is between 2 and 3.

   C. The product is more than $\frac{10}{3}$.

   D. The product is between $\frac{7}{2}$ and 4.5.

FSA Subject Test Mathematics Grade 5

7) A shirt costing $160 is discounted 7%. Which of the following expressions can be used to find the selling price of the shirt?

   A. (160) (0.07)

   B. (160) – 160 (0.17)

   C. (160) (0.07) – (150) (0.70)

   D. (160) (0.93)

8) Which of the following angles is acute?

   A. 110 Degrees

   B. 125 Degrees

   C. 270 Degrees

   D. 75 Degrees

9) If A = 11, then which of the following equations are correct?

   A. A + 30 = 330

   B. A ÷ 30 = 330

   C. 30 × A = 330

   D. A – 30 = 330

10) The perimeter of the trapezoid below is 60. What is its area?

    A. 190

    B. 220

    C. 180

    D. 210

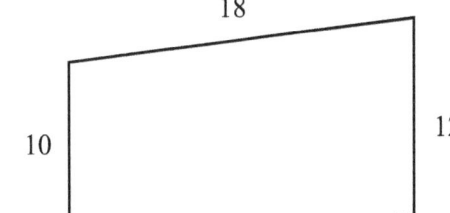

FSA Subject Test Mathematics Grade 5

11) In a bag, there are 108 cards. Of these cards, 12 cards are white. What fraction of the cards are white?

A. $\frac{1}{9}$

B. $\frac{2}{9}$

C. $\frac{1}{12}$

D. $\frac{5}{12}$

12) A rope weighs 600 grams per meter of length. What is the weight in kilograms of 3.8 meters of this rope? (1 kilograms = 1,000 grams)

A. 0.0228

B. 0.228

C. 2.28

D. 2,280

13) Lily and Ella are in a pancake–eating contest. Lily can eat four pancakes per minute, while Ella can eat $3\frac{1}{2}$ pancakes per minute. How many total pancakes can they eat in 8 minutes?

A. 50 Pancakes

B. 58 Pancakes

C. 48 Pancakes

D. 60 Pancakes

FSA Subject Test Mathematics Grade 5

14) How many $\frac{1}{10}$ cup servings are in a package of cheese that contains $3\frac{1}{5}$ cups altogether?

A. $\frac{16}{5}$

B. 32

C. 28

D. $\frac{9}{10}$

15) With what number must 7.551574 be multiplied in order to obtain the number 755,157.4?

A. 100

B. 1,000,000

C. 1,000

D. 100,000

16) What is the volume of this box?

A. 63 cm³

B. 90 cm³

C. 240 cm³

D. 216 cm³

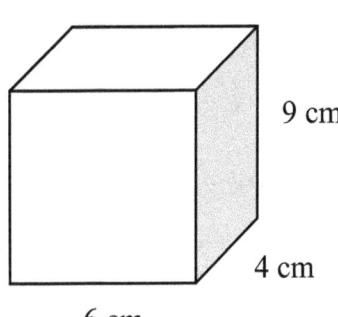

9 cm
4 cm
6 cm

Session 2

❖ **Calculators are NOT permitted for this practice test.**

❖ **Time for Session 1: 80 Minutes**

FSA Subject Test Mathematics Grade 5

17) 18 is What percent of 60?

   A. 80 %

   B. 40 %

   C. 30 %

   D. 100 %

18) The area of a circle is 169π. What is the circumference of the circle?

   A. 13π

   B. 26π

   C. 462π

   D. 56π

19) The distance between cities A and B is approximately 1,260 miles. If Alice drives an average of 45 miles per hour, how many hours will it take Alice to drive from city A to city B?

   A. Approximately 35 Hours

   B. Approximately 28 Hours

   C. Approximately 40 Hours

   D. Approximately 19 Hours

20) 4 yards 9 feet and 24 inches equals to how many inches?

   A. 176

   B. 362

   C. 276

   D. 167

FSA Subject Test Mathematics Grade 5

21) Which expression has a value of – 16?

   A. $9 - (-8) + (-33)$

   B. $2 + (-2) \times (-12)$

   C. $-3 \times (-4) + (-3) \times (-6)$

   D. $(-8) \times (-3) + 4$

22) Solve. $\frac{7}{9} \times \frac{5}{7} =$

   A. $\frac{5}{9}$

   B. $\frac{1}{9}$

   C. $\frac{2}{7}$

   D. $\frac{1}{7}$

23) The length of a rectangle is $\frac{8}{15}$ of inches and the width of the rectangle is $\frac{3}{32}$ of inches. What is the area of that rectangle?

   A. $\frac{1}{15}$

   B. $\frac{1}{20}$

   C. $\frac{11}{47}$

   D. $\frac{15}{112}$

FSA Subject Test Mathematics Grade 5

24) ABC Corporation earned only $120,000 during the previous year, three– second only of the management's predicted income. How much earning did the management predict?

   A. $100,000

   B. $160,000

   C. $80,000

   D. $90,000

25) How many square feet of tile is needed for 19 feet to 19 feet room?

   A. 76 Square Feet

   B. 225 Square Feet

   C. 169 Square Feet

   D. 361 Square Feet

26) Solve. $\frac{3}{7} + \frac{2}{3} - \frac{2}{21} =$

   A. $\frac{4}{7}$

   B. $\frac{7}{10}$

   C. 1

   D. 2

FSA Subject Test Mathematics Grade 5

27) Of the 3,000 videos available for rent at a certain video store, 900 are comedies. What percent of the videos are comedies?

A. $3\frac{1}{30}$ %

B. 3%

C. 15%

D. 30%

28) How many 7 × 7 squares can fit inside a rectangle with a height of 98 and width of 25?

A. 50

B. 79

C. 110

D. 80

29) William keeps track of the length of each fish that he catches. Following are the lengths in inches of the fish that he caught one day: 19, 14, 7, 13, 7, 9, 19

What is the median fish length that William caught that day?

A. 14 Inches

B. 13.5 Inches

C. 75 Inches

D. 13 Inches

FSA Subject Test Mathematics Grade 5

30) $19 + [6 \times 5] \div 3 =?$

A. 21

B. 29

C. 33

D. 26

"This is the end of Practice Test 1."

FSA Subject Test Mathematics Grade 5

FSA Subject Test Mathematics Grade 5

The Florida Standards Assessments

FSA Practice Test 2

Mathematics

GRADE 5

Released *Month Year*

Session 1

❖ **Calculators are NOT permitted for this practice test.**

❖ **Time for Session 1: 80 Minutes**

FSA Subject Test Mathematics Grade 5

1) Jack added 11 to the product of 15 and 18. What is this sum?

   A. 87

   B. 178

   C. 281

   D. 677

2) Joe makes $7.25 per hour at his work. If he works 8 hours, how much money will he earn?

   A. $42

   B. $38

   C. $48

   D. $58

3) What is the value of $7 - 2\frac{2}{3}$?

   A. $\frac{23}{3}$

   B. $4\frac{1}{3}$

   C. $-\frac{2}{3}$

   D. $3\frac{2}{3}$

FSA Subject Test Mathematics Grade 5

4) The bride and groom invited 120 guests for their wedding. 102 guests arrived. What percent of the guest list was not present?

   A. 1.50%

   B. 18%

   C. 180%

   D. 15%

5) Frank wants to compare these two measurements.

   56.067 kg ☐ 56,076 g

   Which symbol should he use?

   A. <

   B. ≠

   C. =

   D. >

6) Aria was hired to teach seven identical 5th grade math courses, which entailed being present in the classroom 21 hours altogether. At $24 per class hour, how much did Aria earn for teaching one course?

   A. $36

   B. $72

   C. $144

   D. $360

FSA Subject Test Mathematics Grade 5

7) In a classroom of 80 students, 48 are male. What percentage of the class is female?

   A. 30%

   B. 25%

   C. 40%

   D. 55%

8) In a party, 13 soft drinks are required for every 18 guests. If there are 216 guests, how many soft drinks are required?

   A. 48

   B. 68

   C. 156

   D. 188

9) You are asked to chart the temperature during an 8–hour period to give the average. These are your results:

   7 am: 3 degrees   11 am: 27 degrees

   8 am: 7 degrees   12 pm: 30 degrees

   9 am: 11 degrees   1 pm: 31 degrees

   10 am: 25 degrees   2 pm: 26 degrees

   What is the average temperature?

   A. 20

   B. 34

   C. 18

   D. 25

FSA Subject Test Mathematics Grade 5

10) While at work, Emma checks her email once every 45 minutes. In 15 hours, how many times does she check her email?

   A. 25 Times

   B. 20 Times

   C. 10 Times

   D. 15 Times

11) In a classroom of 70 students, 28 are male. About what percentage of the class is female?

   A. 70%

   B. 45%

   C. 60%

   D. 30%

12) A florist has 608 flowers. How many full bouquets of 16 flowers can he make?

   A. 42

   B. 26

   C. 28

   D. 38

13) What is 5,287.68925 rounded to the nearest tenth?

   A. 5,287.689

   B. 5,287.7

   C. 5,287

   D. 5,287.69

FSA Subject Test Mathematics Grade 5

14) If a rectangular swimming pool has a perimeter of 84 feet and it is 24 feet wide, what is its area?

   A. 682 square feet

   B. 820 square feet

   C. 1,430 square feet

   D. 432 square feet

15) What is the volume of the following rectangle prism?

   A. 88 ft^3

   B. 188 ft^3

   C. 116 ft^3

   D. 312 ft^3

16) A circle has a diameter of 14 inches. What is its approximate circumference?

   ($\pi = 3.14$)

   A. 21.45 inches

   B. 43.96 inches

   C. 65.48 inches

   D. 32.54 inches

FSA Subject Test Mathematics Grade 5

Session 2

❖ **Calculators are NOT permitted for this practice test.**

❖ **Time for Session 1: 80 Minutes**

FSA Subject Test Mathematics Grade 5

17) How long is the line segment shown on the number line below?

A. 12

B. 5

C. 8

D. 14

18) If $x = -5$, which equation is true?

A. $x(3x + 7) = 45$

B. $4(8 - x) = 52$

C. $6(2x + 12) = 15$

D. $3x - 20 = -38$

19) Peter traveled 360 miles in 4 hours and Jason traveled 560 miles in 7 hours. What is the ratio of the average speed of Peter to average speed of Jason?

A. 9: 8

B. 7: 5

C. 9: 7

D. 7: 4

FSA Subject Test Mathematics Grade 5

20) A woman owns a dog walking business. If 4 workers can walk 16 dogs, how many dogs can 7 workers walk?

    A. 56

    B. 28

    C. 14

    D. 32

21) Which list shows the fractions listed in order from least to greatest?

$$\frac{1}{4}, \frac{1}{18}, \frac{1}{9}, \frac{1}{14}$$

   A. $\frac{1}{14}, \frac{1}{4}, \frac{1}{18}, \frac{1}{9}$

   B. $\frac{1}{9}, \frac{1}{18}, \frac{1}{4}, \frac{1}{14}$

   C. $\frac{1}{4}, \frac{1}{9}, \frac{1}{14}, \frac{1}{18}$

   D. $\frac{1}{18}, \frac{1}{14}, \frac{1}{9}, \frac{1}{4}$

22) What are the coordinates of the intersection of x-axis and the y-axis on a coordinate plane?

    A. $(0, -5)$

    B. $(5, 0)$

    C. $(0, 0)$

    D. $(0, 5)$

FSA Subject Test Mathematics Grade 5

23) In a triangle ABC the measure of angle ACB is 52° and the measure of angle CAB is 76°. What is the measure of angle ABC?

   A. 112

   B. 52

   C. 85

   D. 68

24) David's motorcycle stalled at the beach and he called the towing company. They charged him $ 1.95 per mile for the first 18 miles and then $2.25 per mile for each mile over 18. David was 23 miles from the motorcycle repair shop. How much was David's towing bill?

   A. $39.65

   B. $106.65

   C. $46.35

   D. $56.35

25) A car uses 9 gallons of gas to travel 459 miles. How many miles per gallon does the car get?

   A. 57 miles per gallon

   B. 41 miles per gallon

   C. 51 miles per gallon

   D. 47 miles per gallon

FSA Subject Test Mathematics Grade 5

26) Nine out of 54 students had to go to summer school. What is the ratio of students who did not have to go to summer school expressed, in its lowest terms?

    A. $\frac{5}{6}$

    B. $\frac{1}{5}$

    C. $\frac{1}{9}$

    D. $\frac{5}{9}$

27) A steak dinner at a restaurant costs $11.5. If a man buys a steak dinner for himself and 7 friends, what will the total cost be?

    A. $54

    B. $92

    C. $48

    D. $106

28) If 5 garbage trucks can collect the trash of 35 homes in a day. How many trucks are needed to collect in 77 houses?

    A. 8

    B. 12

    C. 9

    D. 11

FSA Subject Test Mathematics Grade 5

29) A barista averages making 8 cups of coffee per hour. At this rate, how many hours will it take until she's made 440 cups of coffee?

   A. 50

   B. 55

   C. 45

   D. 60

30) Ava needs $\frac{1}{9}$ f an ounce of salt to make 1 cup of dip for fries. How many cups of dip will she be able to make if she has 57 ounces of salt?

   A. 98

   B. 182

   C. 624

   D. 513

"This is the end of Practice Test 2."

FSA Subject Test Mathematics Grade 5

Chapter 14 : Answers and Explanations

FSA Practice Tests

Answer Key

✻ Now, it's time to review your results to see where you went wrong and what areas you need to improve!

Practice Test - 1						Practice Test - 2					
1	C	11	A	21	A	1	C	11	C	21	D
2	A	12	C	22	A	2	D	12	D	22	C
3	A	13	D	23	B	3	B	13	B	23	B
4	B	14	B	24	C	4	D	14	D	24	C
5	D	15	D	25	D	5	A	15	D	25	C
6	B	16	D	26	C	6	B	16	B	26	A
7	D	17	C	27	D	7	C	17	A	27	B
8	D	18	B	28	A	8	C	18	B	28	D
9	C	19	B	29	D	9	A	19	A	29	B
10	B	20	C	30	B	10	B	20	B	30	D

FSA Subject Test Mathematics Grade 5

Practice Test 1
FSA - Mathematics
Answers and Explanations

1) Answer: C.

To find the answer, subtract 49,895, from 50,993.

$50,993 - 49,895 = 1,098 \ miles$

2) Answer: A.

Let x be the new price after discount.

$x = 60 \times (100 - 25)\% = 60 \times 75\% = 60 \times 0.75 = 45 \Rightarrow x = \$45$

3) Answer: A.

Use area of rectangle formula.

Area of a rectangle $= width \times length \Rightarrow D = w \times l \Rightarrow w = \frac{D}{l} = \frac{D}{33}$

4) Answer: B.

6 eggs for 1 cake. Therefore, 540 eggs can be used for $(540 \div 6)$ 90 cakes.

5) Answer: D.

To list the fractions from least to greatest, you can convert the fractions to decimal.

$\frac{5}{8} = 0.625; \frac{8}{10} = 0.8; \frac{1}{10} = 0.1; \frac{5}{9} = 0.556; \frac{3}{7} = 0.428$

$\frac{1}{10} = 0.1, \frac{3}{7} = 0.428, \frac{5}{9} = 0.556, \frac{5}{8} = 0.625, \frac{8}{10} = 0.8$

Option D shows the fractions in order from least to greatest.

6) Answer: B.

6 multiplied by $\frac{3}{8} = \frac{18}{8} = 2.25$, therefore, only choice B is correct.

7) Answer: D.

To find the selling price, multiply the price by (100% – rate of discount).

Then: $(160)(100\% - 7\%) = (160)(0.93) = 148.8$

8) Answer: D.

An acute angle is any angle smaller than 90 degrees.

From the options provided, only D (75 degrees) is smaller than 90.

WWW.MathNotion.Com

FSA Subject Test Mathematics Grade 5

9) Answer: C.

Plug in 11 for A in the equations. Only option A works.

$A \times 30 = 330$; $11 \times 30 = 330$

10) Answer: B.

First, find the missing side of the trapezoid. The perimeter of the trapezoid below is 60. Therefore, the missing side of the trapezoid (its height) is:

$60 - 10 - 12 - 18 = 60 - 40 = 20$

Area of a trapezoid: $A = \frac{1}{2} h (b1 + b2)$

$= \frac{1}{2} (20) (10 + 12) = 220$

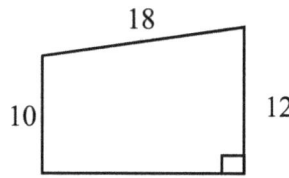

11) Answer: A.

There are 108 cards in the bag and 12 of them are white. Then, 12 out of 108 cards are white. You can write this as: $\frac{12}{108}$. To simplify this fraction, divide both numerator and denominator by 12. Then: $\frac{12}{108} = \frac{1}{9}$

12) Answer: C.

1 meter of the rope = 600 grams

3.8 meter of the rope = $3.8 \times 600 = 2,280$ grams = 2.28 kilograms

13) Answer: D.

Lily eats 4 pancakes in 1 minute ⇒ Lily eats 4×8 pancakes in 8 minutes (32).

Ella eats $3\frac{1}{2}$ pancakes in 1 minute ⇒ Ella eats $3\frac{1}{2} \times 8$ pancakes in 8 minutes ($\frac{56}{2} = 28$).

In total Lily and Ella eat $32 + 28 = 60$ pancakes in 8 minutes.

14) Answer: B.

To solve this problem, divide $3\frac{1}{5}$ by $\frac{1}{10}$.

$3\frac{1}{5} \div \frac{1}{10} = \frac{16}{5} \div \frac{1}{10} = \frac{16}{5} \times \frac{10}{1} = 32$

15) Answer: D.

The question is that number 755,157.4 is how many times of number 7.551574. The answer is 100,000.

FSA Subject Test Mathematics Grade 5

16) Answer: D.

Use volume of cube formula.

Volume= $length \times width \times height \Rightarrow V = 6 \times 4 \times 9 \Rightarrow V = 216\ cm^3$

17) Answer: C.

Use percent formula: $part = \frac{percent}{100} \times whole$

$18 = \frac{percent}{100} \times 60 \Rightarrow 18 = \frac{percent \times 60}{100} \Rightarrow 18 = \frac{percent \times 6}{10}$, multiply both sides by 10.

$180 = percent \times 6$, divide both sides by 6. $\Rightarrow 30$ percent

18) Answer: B.

Use area and circumference of circle formula.

Area of a circle = $\pi r^2 \Rightarrow 169\pi = \pi r^2 \Rightarrow r = 13$

Circumference of a circle = $2\pi r \Rightarrow C = 2 \times 13 \times \pi \Rightarrow C = 26\pi$

19) Answer: B.

Alice drives 45 miles in one hour. Therefore, she drives 1,260 miles in about (1,260 ÷ 45) 28 hours.

20) Answer: C.

4 yards = 4 × 36 = 144 inches, 9 feet = 9 × 12 = 108 inches

4 yards 9 feet and 24 inches = 144 inches + 108 inches + 24 inches = 276 inches

21) Answer: A.

Simplify each option provided using order of operations rules.

A. $9 - (-8) + (-33) = 9 + 8 - 33 = -16$

B. $2 + (-2) \times (-12) = 2 + 24 = 26$

C. $-3 \times (-4) + (-3) \times (-6) = 12 + 18 = 30$

D. $(-8) \times (-3) + 4 = 24 + 4 = 28$

Only option A is -16.

22) Answer: A.

$\frac{7}{9} \times \frac{5}{7} = \frac{7 \times 5}{9 \times 7} = \frac{35}{63} = \frac{5}{9}$

FSA Subject Test Mathematics Grade 5

23) Answer: B.

Use area of rectangle formula.

Area= length × width $\Rightarrow A = \frac{8}{15} \times \frac{3}{32} \Rightarrow A = \frac{1}{20}$ inches

24) Answer: C.

ABC Corporation's income = $\frac{3}{2}$ management's predicted income.

$120,000 = $\frac{3}{2}$ management's predicted income

management's predicted income = $120,000 × $\frac{2}{3}$ = $80,000

25) Answer: D.

Find the area of the room which is a square. Use area of square formula.

$S = a^2 \Rightarrow S = 19\ feet \times 19\ feet = 361$ square feet

26) Answer: C.

$\frac{3}{7} + \frac{2}{3} - \frac{2}{21} = \frac{(3\times3)+(2\times7)-(2\times1)}{21} = \frac{21}{21} = 1$

27) Answer: D.

Use percent formula: part $= \frac{percent}{100} \times$ whole

$900 = \frac{percent}{100} \times 3{,}000 \Rightarrow 900 =$ percent $\times 30 \Rightarrow$ percent $= 30$

28) Answer: A.

Use area of rectangle formula. $A = a \times b \Rightarrow A = 98 \times 25 \Rightarrow A = 2{,}450$

Divide the area by 49 (7 × 7 = 49 squares) to find the number of squares needed.

2,450 ÷ 49 = 50

29) Answer: D.

Write the numbers in order: 7, 7, 9, 13, 14, 19, 19

Median is the number in the middle. Therefore, the median is 13.

30) Answer: B.

Use PEMDAS (order of operation): 19 + [6 × 5] ÷ 3 = 19 + (30) ÷ 3 = 19 + (30 ÷ 3) = 19 + 10 = 29

Practice Test 2
FSA - Mathematics
Answers and Explanations

1) Answer: C.

11 + (15 × 18) = 11 + 270 = 281

2) Answer: D.

1 hour: $7.25

8 hours: 8 × $7.25 = $58

3) Answer: B.

$7 - 2\frac{2}{3} = \frac{21}{3} - \frac{8}{3} = \frac{13}{3} = 4\frac{1}{3}$

4) Answer: D.

The number of guests that are not present are: (120 – 102) 18 out of 120 = $\frac{18}{120}$

Change the fraction to percent: $\frac{18}{120} \times 100\% = 15\%$

5) Answer: A.

Each kilogram is 1,000 grams. 56,076 grams = $\frac{56,076}{1,000}$ = 56.076 kilograms. Therefore, 56,076 grams are greater than the 56.067 kg.

6) Answer: B.

Aria teaches 21 hours for seven identical courses. Therefore, she teaches 3 hours for each course. Aria earns $24 per hour. Therefore, she earned $72 (3 × 24) for each course.

7) Answer: C.

The number of female students in the class is: (80 – 48) 32 out of 80 = $\frac{32}{80}$

Change the fraction to percent: $\frac{32}{80} \times 100\% = 40\%$

8) Answer: C.

Write a proportion and solve.

$\frac{13 \text{ soft drinks}}{18 \text{ guests}} = \frac{x}{216 \text{ guests}} \Rightarrow x = \frac{216 \times 13}{18} \Rightarrow x = 156$

FSA Subject Test Mathematics Grade 5

9) Answer: A.

average (mean) = $\dfrac{\text{sum of terms}}{\text{number of terms}}$ ⇒ average = $\dfrac{3+7+11+25+27+30+31+26}{8}$ ⇒ average = 20

10) Answer: B.

Every 45 minutes Emma checks her email.

In 15 hours (900 minutes), Emma checks her email (900 ÷ 45) 20 times.

11) Answer: C.

There are 70 students in the class. 28 of the are male and 42 of them are female. 42 out of 70 are female. Then:

$\dfrac{42}{70} = \dfrac{x}{100}$ → 4,200 = 70x → x = 42,00 ÷ 70 = 60%

12) Answer: D.

Divide the number flowers by 16: 608 ÷ 16 = 38

13) Answer: B.

Rounding decimals is similar to rounding other numbers. If the hundredths and thousandths places of a decimal is sixty-eight or less, they are dropped, and the tenths place does not change. For example, rounding 0.724 to the nearest tenth would give 0.7. Therefore, 5,287.68925 rounded to the nearest tenth is 5,287.7.

14) Answer: D.

Perimeter of rectangle formula:

$P = 2\,(length + width)$ ⇒ 84 = 2 (l + 24) ⇒ l = 18

Area of rectangle formula: $A = length \times width$ ⇒ A = 24 × 18 ⇒ A = 432

15) Answer: D.

Use volume of rectangle prism formula.

$V = length \times width \times height$ ⇒ V = 13 × 3 × 8 ⇒ V = 312

16) Answer: B.

The diameter of the circle is 14 inches. Therefore, the radius of the circle is 7 inches.

Use circumference of circle formula. $C = 2\pi r$ ⇒ C = 2 × 3.14 × 7 ⇒ C = 43.96

17) Answer: A.

The line segment is from 7 to −5. Therefore, the line is 12 units.

7 − (−5) = 7 + 5 = 12

WWW.MathNotion.Com

FSA Subject Test Mathematics Grade 5

18) Answer: B.

Plug in $x = -5$ in each equation.

A. $x(3x + 7) = 45 \rightarrow (-5)(3(-5) + 7) = (-5) \times (-15 + 7) = 40$

B. $4(8 - x) = 52 \rightarrow 4(8 - (-5)) = 4(13) = 52$

C. $6(2x + 12) = 15 \rightarrow 6(2(-5) + 12) = 6(-10 + 12) = 12$

D. $3x - 20 = -38 \rightarrow 3(-5) - 20 = -15 - 20 = -35$

Only option B.

19) Answer: A.

Peter's speed = $\frac{360}{4} = 90$

Jason's speed = $\frac{560}{7} = 80$

$\frac{The\ average\ speed\ of\ peter}{The\ average\ speed\ of\ Jason} = \frac{90}{80}$ equals to: $\frac{9}{8}$ or 9: 8

20) Answer: B.

4 workers can walk 16 dogs ⇒ 1 workers can walk 4 dogs.

7 workers can walk (7 × 4) 28 dogs.

21) Answer: D.

In fractions, when denominators increase, the value of fractions decrease and as much as numerators increase, the value of fractions increase. Therefore, the least one of this list is: $\frac{1}{18}$ and the greatest one of this list is: $\frac{1}{4}$

22) Answer: C.

The horizontal axis in the coordinate plane is called the $x - axis$. The vertical axis is called the $y - axis$. The point at which the two axes intersect is called the origin. The origin is at 0 on the $x - axis$ and 0 on the $y - axis$.

23) Answer: B.

All angles in every triangle add up to 180°. Let x be the angle ABC. Then: 180 = 52° + 76° + x ⇒ $x = 52°$

24) Answer: C.

$1.95 per mile for the first 18 miles. Therefore, the cost for the first 18 miles is:

FSA Subject Test Mathematics Grade 5

18 × $1.95 = $35.1

$2.25 per mile for each mile over 18, therefore, 5 miles over 18 miles cost:

5 × $2.25 = $11.25

In total, David pays: $35.1+ $11.25= $46.35

25) Answer: C.

Write a proportion and solve.

9 gallons: 459 miles ⇒ 1gallon: 459 ÷ 9 = 51 miles

26) Answer: A.

The students that had to go to summer school is 9 out of 54 $= \frac{9}{54} = \frac{1}{6}$

Therefore $\frac{5}{6}$ students did not have to go to summer school.

27) Answer: B.

8 steak dinners = 8 × $11.5 = $92

28) Answer: D.

5 garbage trucks can collect the trash of 35 homes. Then, one garbage truck can collect the trash of 7 homes.

To collect trash of 77 houses, 11 (77 ÷ 7) garbage trucks are required.

29) Answer: B.

8 cups: 1 hour

440 cups: 440 ÷ 8 = 55 hours

30) Answer: D.

Write a proportion and solve.

$\frac{\frac{1}{9}}{57} = \frac{1}{x} \Rightarrow x = 57 \times 9 = 513$

"End"

www.ingramcontent.com/pod-product-compliance
Lightning Source LLC
Chambersburg PA
CBHW080440110426
42743CB00016B/3222